# PATENT FUNDAMENTALS

## for SCIENTISTS and ENGINEERS

### Second Edition

# PATENT FUNDAMENTALS
## for SCIENTISTS and ENGINEERS
## Second Edition

Thomas T. Gordon
Arthur S. Cookfair

## LEWIS PUBLISHERS
Boca Raton  London  New York  Washington, D.C.

*Library of Congress Cataloging-in-Publication Data*

Gordon, Thomas T.
   Patent fundamentals for scientists and engineers / Thomas T. Gordon, Arthur S.
Cookfair.— 2nd ed.
      p.   cm.
   Includes bibliographical references and index.
   ISBN 1-56670-517-7 (alk. paper)
   1. Patents.   2. Patents — United States.   I. Cookfair, Arthur S.   II. Title.

T339.G67 2000
608.773 — dc21                                                                                      99-056999
                                                                                                        CIP

© 2000 by CRC Press LLC
Lewis Publishers is an imprint of CRC Press LLC

No claim to original U.S. Government works
International Standard Book Number 1-56670-517-7
Library of Congress Card Number 99-056999
Printed in the United States of America  1  2  3  4  5  6  7  8  9  0
Printed on acid-free paper

# About the Authors

**Thomas T. Gordon.** After graduation from college with a B.S. in chemistry, Mr. Gordon worked as a chemist in the pharmaceutical and polymer industries. He did graduate work in organic chemistry and studied law, receiving his J.D. from St. Louis University.

Mr. Gordon has over 30 years' experience in the patent field, working in various areas of patent preparation, searching, and prosecution. He has conducted patent licensing programs in the U.S. and Europe, and has worked for major industrial corporations. He is currently in private patent practice in the Arlington, Virginia, area.

**Arthur S. Cookfair** is a patent agent with more than 25 years' experience in patent law in both corporate and private practice. He has, in addition, served as a patent examiner in the U.S. Patent and Trademark Office and is the author of numerous publications on patents and inventions.

Based on his undergraduate work in chemistry and geoscience, and graduate work in science and education, Dr. Cookfair has had a varied career as a chemist, educator, and patent practitioner. He has lectured and taught extensively on science and patent law in this country and abroad and has been the recipient of a Fulbright grant.

# Contents

# Preface to the Second Edition

Although the first patent law was passed in 1790, patents have been a mysterious subject filled with misconceptions and ideas. The patent professionals who understood the patent laws and their benefits were a limited group of people, and their knowledge was not transferred to the broad group of experimenters, scientists, and engineers who generate patentable inventions. The understanding of what a patent is, how it is obtained, and its value is still a mystery.

This edition of *Patent Fundamentals for Scientists and Engineers* will attempt to solve that problem. This book should provide to the independent inventor as well as to members of the scientific and business community — whether a scientist, engineer, supervisor, manager, or senior scientist — an overview of the patent system. The book also provides a guide to assist them in their dealings with the U.S. Patent and Trademark Office, as well as with patent professionals. The patent profession consists of attorneys and agents, whose duties are to evaluate the concept, prepare the application, prosecute the application, and obtain the maximum protection for the inventor. The inventor and the businessman should participate in all these steps.

The patent system is also designed to promote the incentive for others to improve on the invention and enlarge the field of technology. This is a vital aspect of our system and has been very successful in the 200+ years it has been in existence. During those years, there have been changes in the system, but the basic tenets established at its creation still exist. Modifications and new applications are constantly occurring. Those changes will occur in the future, but the system will remain strong and will continue to foster advances in science and technology.

*chapter one*

# Patents as intellectual property

"The Congress shall have the power...to promote the progress of science and useful arts, by securing for limited times to authors and inventors the exclusive rights to their respective writings and discoveries."

*The U.S. Constitution*

In drafting Article I, Section 8, Clause 8 of the Constitution, the framers of that document achieved the dual purpose of establishing the basis for the U.S. patent system as well as the copyright system. The clause sets forth three balanced phrases — "science and the useful arts," "authors and inventors," and "writings and discoveries" — to refer to copyrights and patents, respectively. By focusing on the words, "science," "authors," and "writings," the clause may be read:

"The Congress shall have the power...to promote the progress of science...by securing for limited times to authors...the exclusive rights to their...writings."

These words provide the constitutional basis for our copyright system. (In the 1700s, "science" referred to learning or knowledge.) A shift to the alternate set of words — "useful arts," "inventors," and "discoveries" — provides the basis for the U.S. Patent System. The expression "useful arts" is simply an old-fashioned term for technology — the downstream useful product of science. The patent system applies to technology, not science. The word "discoveries" has been interpreted as meaning "inventions." Thus, the constitutional basis for our patent system is found in the words:

"The Congress shall have the power...to promote the progress of...useful arts, by securing for limited times to...inventors the exclusive right to their...discoveries."

The language not only provides the authority for our patent system, but clearly establishes its purpose: not simply to protect the rights of the inventors but to promote the progress of technology — a broader, societal purpose. Patents and copyrights are both categories that fall within the broader concept of intellectual property. The concept is termed "intellectual" because it applies to products of the mind, and "property" because those products belong to the person whose mental efforts created them. However, the concept of property also denotes ownership and exclusive rights. Since the property is mental, it is apparent that ownership and exclusive rights can be readily maintained only as long as the mental property is not revealed to anyone else. This leads to the dilemma recognized by Thomas Jefferson when he wrote:

> "If nature has made any one thing less susceptible than all others of exclusive property, it is the action of the thinking power called an idea, which an individual may exclusively possess as long as he keeps it to himself..."

Jefferson recognized the problem that is addressed by the laws of intellectual property; that is, how can a person freely and openly put an idea to use, thus divulging it to others, and still retain exclusive rights to the idea?

It has been nearly 200 years since Thomas Jefferson observed the difficulty in protecting an idea by providing exclusive rights to its originator or owner, and the concept of "intellectual property" has only recently become generally recognized. Just 30 years ago, the term was largely unknown, except to a few lawyers. Now it has become a major concern of the governments of the U.S. and other nations and an important consideration in international trade negotiations. Since Jefferson's time, there have evolved, in the U.S. and other nations, bodies of law directed to the protection of that intangible property known as intellectual property (as opposed to tangible property, e.g., real estate).

Current authors now term Article 1, Section 8 the Intellectual Property clause of the Constitution.

In general, intellectual property laws are directed to the protection of rights arising from creative mental activities and are embodied in the laws (both statutory and common law) of trademarks, copyrights, trade secrets, and patents. Each of these categories includes a system of laws designed to protect a particular kind of intellectual property.

## Trademarks

A trademark is a distinctive word or phrase, name, symbol, or device used to identify the source of goods in commerce. A trademark on a product serves to identify the manufacturer or seller and to distinguish it from the products of others. Trademarks are commonly used to identify a brand of merchandise and may be used to promote a company's products and establish a symbol that can be utilized in advertising and marketing. The term "trademark" also encompasses service marks, which identify the source of services rather than goods.

The trademark is not used to identify a product — only the *source* of the product. Thus, others may have the right to manufacture and sell an identical product, but not to use the same trademark (without permission) in connection with the product.

In the U.S., the registration of trademarks is handled by the U.S. Patent and Trademark Office, the same federal agency that handles the granting of patents. There are approximately 900,000 active registered trademarks on the principal register of the U.S. Patent and Trademark Office, with approximately 60,000 new registrations each year.

Unlike patents and copyrights, trademarks can remain in force indefinitely, subject to renewal every 20 years, provided the mark has been continuously in use.

## Copyrights

Copyright laws are directed at the protection of the creative works of authors, artists, and others from unauthorized copying. U.S. copyright laws are constitutionally based and, along with patents, find their basis in Article 1, Section 8 of the Constitution. Although the original provision referred to authors and writings, Congress and the federal courts have interpreted it to include other forms of literary or artistic expression, such as musical compositions, dramatic works, maps, photographs, paintings, sculptures, etc.

The subject matter that can be protected by copyright under present copyright laws includes:

- Literary works
- Musical works
- Dramatic works
- Pantomimes and choreographic works
- Pictorial, graphic, and sculptural works
- Motion pictures and other audiovisual works
- Sound recordings
- Computer programs

A copyright protects the manner of expression of an idea — not the underlying idea itself.

The registration of copyrights is administered by the Copyright Office, a division of the Library of Congress. The term of the protection of a copyright is for the life of the author plus 50 years. When the copyright expires, the protected work can be freely copied or used by anyone.

## Trade secrets

In a competitive industry, confidential information can be a valuable asset. Many businesses rely heavily on the confidentiality or secrecy of information relating to the manufacturing processes, product specifications, product treatment methods, and even customer lists and other business information to give them a competitive edge over others. Such information is often referred to as a "trade secret" and can be protected under the body of common law known as trade secret law. In contrast to patents, trademarks, and copyrights, which are governed by federal statutes, trade secret law falls within the jurisdiction of the states. Although there is no universal agreement on exactly what constitutes a trade secret, the most commonly accepted definition is set forth in the legal reference work *The Restatement of Torts, Section 757*, which states:

> "A trade secret may consist of any formula, pattern, device, or compilation of information which is used in one's business, and which gives him an opportunity to obtain an advantage over competitors who do not know or use it. It may be a formula for a chemical compound, a process of manufacturing, treating, or preserving materials, a pattern or a machine or other device, or a list of customers."

## Patents

Patent laws are the division of intellectual property laws that focuses on the protection of inventors. A patent is a grant by the government, to an inventor, conferring the right to exclude others from making, using, or selling his/her invention for a limited period of time. U.S. patents are granted for a period that begins at the date of issue and ends 20 years from the date that the application was filed.

The subject matter for which patents are granted includes invention of new and useful processes, machines, articles of manufacture, and compositions of matter.

Although trademarks, copyrights, and patents are all means of protecting intellectual property, the subject matter protectable by each is clearly different. However, the subject matter protectable by patents and trade secret laws overlaps considerably, and the ramifications of that overlap will be treated in further detail in a later chapter.

## chapter two

# Patents: history, philosophy, and purpose

"The patent system added the fuel of interest to the fire of genius."

Abraham Lincoln

Patents, in terms of today's thinking, find their roots at least as far back as the 15th century. The city-state of Venice, which dominated the Mediterranean region during the latter part of the Middle Ages, adopted the first patent law. The preamble to that 1474 Venetian Patent Act set forth a philosophy that can be found in all modern patent systems: a recognition that the grant of exclusive rights to an inventor, for a limited time, will encourage others to invent.

It was under the Venetian system that Galileo applied for a patent in 1593 for his invention of a pump to supply water for irrigation. The pump required a single horse for power (one-horsepower pump) and discharged water through 20 spouts. The application was examined and a patent granted decreeing that for a period of 20 years, Galileo should have exclusive rights and that any infringer would forfeit the infringing apparatus and pay a fine of 200 ducats.

The granting of patents was practiced by other European nations, with variations in practice from country to country. Often, the practice was simply a method of rewarding friends of the ruler and was not primarily intended to reward creativity and encourage inventors.

The modern concept of a patent system began to take shape in England during the 16th and 17th centuries, but not without difficulty. Abuses were rampant during that period. Patents were often granted by the Crown for political or financial considerations. Exclusive rights were granted (or sold) to favorites of the court, giving them monopolies in certain commodities or services, often with no invention being required. Finally in 1624, in response to such abuses, the British Parliament adopted the Statute of Monopolies, which outlawed all royal monopolies except patent grants to inventors of new manufactures. The patent term was limited to 14 years and was granted only to the first and true inventor.

Immigrants from England brought the concept of patents to the New World, and the British experience provided a springboard for the establishment of a patent system in America. Prior to the adoption of the U.S. Constitution, patents were granted by the individual colonies or states. Such patents were generally granted by special acts of the legislature. There were no general laws providing for the granting of patents, and the terms of those patents varied greatly. The first of these patents was granted to Samuel Winslow by the Massachusetts General Court in 1641 and related to a "Method of Making Salt." It had a term of 10 years. In 1747, Thomas Darling of Connecticut received a 20-year patent on a process for "Making Glass."

In 1787, when the state delegates assembled in Philadelphia to frame the Constitution, the nation was an agriculturally based country. It had little or no manufacturing facilities and was dependent on other countries to provide the necessary tools and machinery to function and, as a result, did not have a highly mechanical workforce. Many of the leaders of the country realized the need for the machinery being developed in Europe, but did not know how to encourage importation across the Atlantic and help turn the nation into a manufacturing economy.

The Industrial Revolution was still in its infancy in England, and although they had a crude system granting monopoly power to a few, patents as we know them were not yet a factor.

The drafters of the Constitution understood that to convert our nation from agriculture to manufacturing and to provide the tools and equipment for progress and national security, a new approach had to be taken. Patents were an approach; so a clause was inserted in our Constitution that granted authority to Congress to establish a patent system. The concept of granting limited-term rights to inventors was not original to the framers of the Constitution, but was known and practiced in Europe.

When the first patent law was enacted by Congress in 1790, there were many suggestions as to whether a reward should be given to the inventors. It was suggested that several acres of land be the reward, as well as monetary sums. None were adopted, but the tenor of the feelings at that time was that a patent was a mechanism for rewarding legitimate inventors and protecting their rights.

Although the Constitution stated that the patent law should "promote the progress of science and useful arts," the patent laws were not being directed to that end. Instead, they were being directed toward rewarding the inventor.

Article I, Section 8, Clause 8 recognizes the value of patents as a stimulus to industry in a developing nation and provides the authority for the U.S. Patent System. That clause not only provides the authority for the patent system, but clearly establishes the purpose: "to promote the progress of [the] useful arts." Thus, the basic purpose of U.S. patent laws is to not simply protect inventors' rights to their inventions; it has a broader societal purpose:

to promote the progress of technology. This purpose has been recently confirmed in a statement by the Court:

> "The patent law is directed to the public purposes of fostering technological progress, investment in research and development, capital formation, entrepreneurship, innovation, natural strength, and international competitiveness."

> Hilton Davis v. Warner-Jenkinson,<br>35USPQ2nd, pp1660.

The purpose has been well-served. The 200-year history of our nation is interwoven with invention, patents, and technological progress.

Congress, in its second session, passed a bill titled "Act to Promote the Progress of the Useful Arts." The bill was signed by President George Washington on April 10, 1790, and the United States Patent System became a reality. In that same year, Samuel Hopkins of Philadelphia received the first U.S. patent. His invention was a chemical process for making potash. Hopkins' patent served to confirm the purpose of the new patent system: to promote the progress of the useful arts. In 1790, potash, an impure form of potassium carbonate, was America's first industrial chemical. His patent was important, not only because it was the first U.S. patent, but also because it provided the vehicle for licensing this new technology to U. S potash producers. The patent grant was signed by George Washington, President of the United States, and also carried the signatures of Edmund Randolph, Attorney General, and Thomas Jefferson, Secretary of State (see Figure 2.1).

Patents are territorial. Patents granted in each nation are enforceable only within the territory of that nation. As a result, a separate patent on the same invention must be obtained in each country in which the owner of the invention wants to protect it. (See Chapter 13 for a more detailed description of worldwide patents.)

## *What a patent is*

The answer to the question, "What is a patent?" can vary greatly, depending on one's perspective. It can be viewed as a technical publication or as a legal document. It can be a valuable asset to the owner or a nuisance to a competitor. It can be a business property that might be used either offensively or defensively. The definition of a patent is set forth in the law, Title 35 of the U.S. Code, Section 154 (35 USC 154):

> "...a grant to the patentee...of the right to exclude others from making, using, offering for sale, or selling the invention throughout the United States...for a term beginning on the date on which the patent issues and

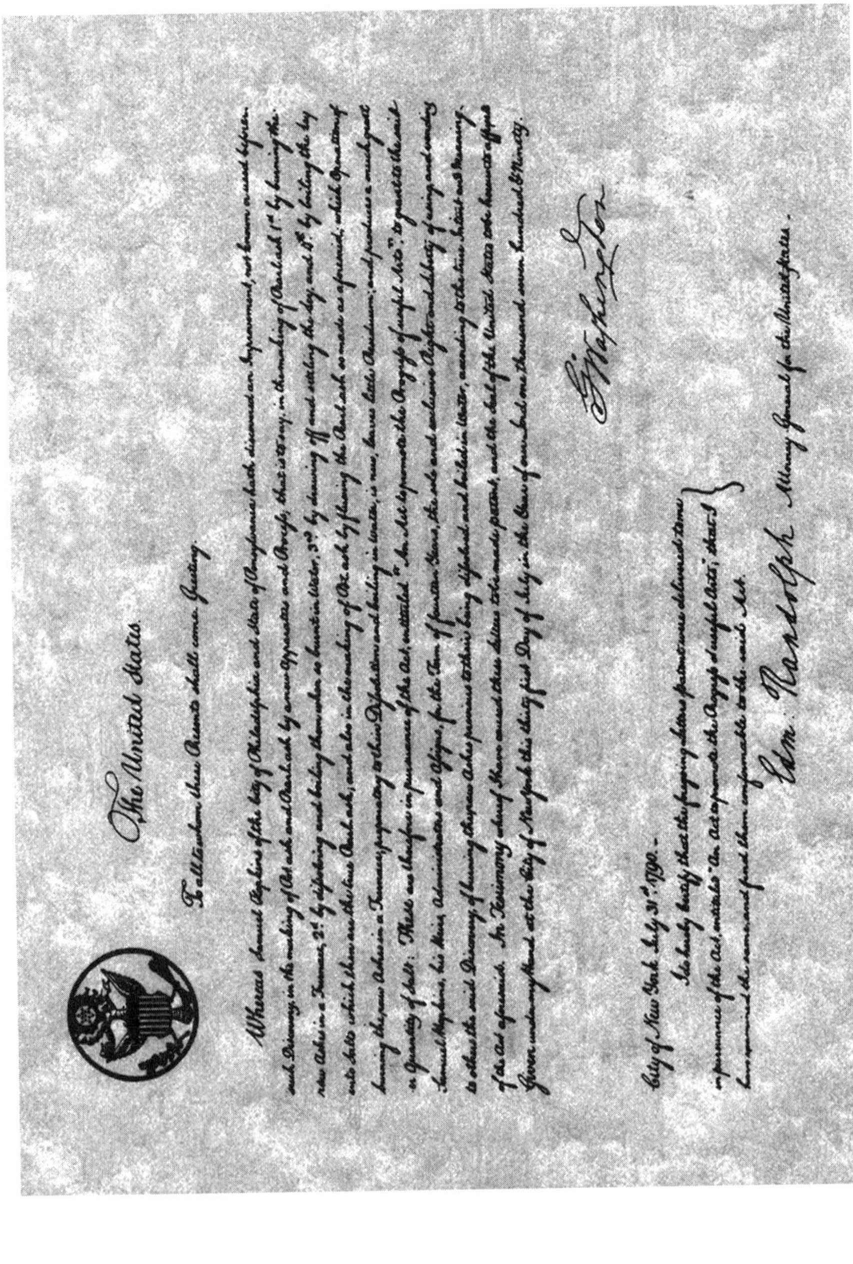

Figure 2.1   First U.S. patent issued.

> ending 20 years from the date on which the application
> was filed..."

In this definition, the word "exclude" is key to an understanding of patents. The patent grant provides the patentee with the right to *exclude* others from practicing the invention.

There are three basic characteristics commonly attributed to patents:

1. *The contractual characteristic*: In viewing the patent as a contract, the focus is on the contractual exchange between the two parties — the inventor and the government. The inventor gives a full disclosure of the invention and how to make and use it, so that when the patent expires, the public will be in possession of the invention and will be able to use the invention freely. In return, the government gives the inventor the right to exclude others from making, using, or selling his/her invention for a limited period of time.
2. *The property characteristic*: Patents have a right of ownership similar to real property that can be transferred in whole or in part. In a manner analogous to real estate, a patent can be sold, or it can be "rented" (licensed) and rent collected for its use (royalties).
3. *The monopoly characteristic*: The "monopoly" associated with patents is limited in time (20 years) and negative in nature, (i.e., exclusionary).

## What a patent is not

A patent is *not* a license to make, use, or sell an invention. It does *not* give the patentee the right to practice his/her invention.

The most common misconception concerning patents is that a patent gives one the right to practice his/her invention. In fact, the patent merely grants the patentee the right to *exclude others* from practicing the invention. *The inventor's right to practice his/her invention is independent of the patent.*

It is not uncommon for an inventor to be granted a patent and not have the right to practice the invention. This happens, for example, when a patented invention is directed to subject matter that falls within the exclusive rights of someone else's earlier, broader patent. The earlier, broader patent is said to dominate, and the first patentee may have the right to stop the second patentee from practicing his/her own invention. Neither patent grants the positive right to practice the invention — only the negative right to exclude others.

To more clearly illustrate the point, consider the following hypothetical examples.

### Example 1: The invention of the knife

Assuming that knives were not known, someone (inventor A) invents the knife and obtains a patent on the invention, describing and defining it as "a tool or weapon comprising a handle and a blade having a cutting edge." The patent gives inventor A the right to exclude others from making, using, or selling knives. Inventor A also has the right to make use of and sell knives. But remember, that right did not come from the patent grant. Inventor A already had that right before the patent grant. The patent only granted the right to exclude others.

The second phase of this scenario involves the invention of the folding knife (i.e., the pocket knife). Inventor B greatly admires the invention of the knife but recognizes its deficiencies. In particular, when the knife is carried in a pocket, the sharp edge tends to cut a hole in the pocket. Inventor B solves the problem by creating a new invention — a folding knife. The inventor is rewarded with a patent for the invention of "a knife comprising a handle and a blade that may be folded into the handle to shield the blade when not in use."

The outcome: inventor A can continue to make, use, and sell knives, but is excluded by inventor B's patent from making, using, or selling folding knives.

Inventor B is excluded from making, using, or selling *any* knife (including his/her own folding knife) that has a handle and a blade with a cutting edge.

Thus, although inventor B has a patent on the folding knife, he/she does not have the right to make, use, or sell the invention, without permission from inventor A, so long as inventor A's patent is in force. Once inventor A's patent expires, inventor B and everyone else has the right to make, use, and sell knives, but only inventor B has the right to make folding knives (until inventor B's patent expires).

To further illustrate the point, consider the following example involving a chemical invention.

## Example 2: The invention of a new polyester

An inventor at company A (a polymer manufacturer) invents a new polyester that has surprising and unusual mechanical properties. This new polymer is made by a simple process, from extremely inexpensive monomers, and the final polymer can be easily and cheaply fabricated into virtually any shape or form. Furthermore, the final polyester exhibits mechanical properties that permit it to be used to fabricate practically any product used in the building trades. Using simple extrusion, casting, or molding techniques, it can be formed into cheaper and stronger plastic equivalents of wooden planks, concrete blocks, roofing, siding, etc.

Company A obtains a patent on this unusual polyester, establishes a manufacturing facility and a marketing organization, and sets out to revolutionize the construction industry. The new business is an overnight success; soon, houses, stores, barns, and office buildings are built across the country from this new polyester. Then disaster strikes. There are several serious fires and a concerned public calls for stricter fire codes. The new polyester does not meet the fire codes, and company A's research team cannot find a way to make the polyester fire-retardant without destroying its unique mechanical properties.

Company A still has the exclusive right to manufacture the polyester, but its primary market has collapsed.

Company B is a small research-oriented chemical company that specializes in developing new polymer additives. Company B's research team experiments with company A's polyester and discovers a new compound (X) that, when added to the polyester in very small amounts, will render it fire-retardant with no detrimental effect on mechanical properties. Company B obtains a patent on this new composition: a fire-retardant composition comprising the polyester and a fire-retardant additive (X).

Company B now has a formulation that will meet the new fire codes but cannot market it because their composition would infringe on company A's patent. Company A cannot market the polyester to the construction industry because it does not meet the fire codes, and they cannot solve the problem by adding fire-retardant (X) to their polymer because to do so would infringe on company B's patent. Stalemate?

At this point, if profits are to be gained from their patented inventions (and more important, if society is to benefit from this new technology), companies A and B must cooperate in some mutually beneficial way. Some of the possibilities include:

Company A could sell its patent rights to company B
Company B could sell its patent rights to company A
Company A could license its patent rights to company B
Company B could license its patent rights to company A
Companies A and B could cross-license
Companies A and B could set up a joint venture company

A study of the patents in the area of one's invention is critical to the determination of whether or not an inventor can use his/her own invention.

## chapter three

# The U.S. Patent System

In the early years following the passage of the first patent statute in 1790, the patent system was simple. The population of the U.S. was less than 4 million, and the numbers of inventions were few. The fees for a patent were about $5.00, and the term was 14 years. The responsibility for granting patents was placed with a board consisting of the Secretary of State, the Secretary of War, and the Attorney General. As Secretary of State, Thomas Jefferson was the first administrator of the American Patent System and was, in effect, the first patent examiner. Today, the U.S. Patent Office employs nearly 3000 patent examiners. Although the system has grown enormously in size and complexity since Jefferson's time, the fundamental principles have remained essentially the same.

As Chapter 2 illustrated, it should be stressed that the issuance of a U.S. patent is *not* a grant of an affirmative right to practice the invention covered by the patent. It is a grant of the right to exclude others from making, using, offering to sell, or selling within the U.S. or importing into the U.S., the claimed invention for a period beginning at the date of issue of the patent and ending 20 years from the date on which the application was filed.

The theory behind the patent grant is that the disclosure to the public of an invention, which will be theirs at a later date, will provide great incentive to create other useful inventions. This theory has been extremely successful, as the U.S. Patent System has grown to be the dominant system in the world, and the growth of industry and technology in the U.S. has been a direct result. In more than 200 years since its inception, over 6 million patents have been issued, and the line of applications awaiting action is still very long.

Throughout the years since the formation of the U.S. Patent Office, there have been additions to its function and duties. Today, its function is to administer and issue trademarks as well as patents. Its official name is the U.S. Patent and Trademark Office (hereafter, the PTO). It is a branch of the Department of Commerce and is staffed by more than 5000 employees, of which about half are patent examiners. This staff approves and issues about 1500 to 2000 patents each week.

The U.S. Patent Office issues three types of patents:

1. **Design patents** are granted to protect any new and original design for an article of manufacture.
2. **Plant patents** are granted for the invention or discovery of a distinct and new variety of plant.
3. **Utility patents** are granted for any new and useful process, machine, article of manufacture, or composition of matter.

Examples of each of these different types of patents are illustrated. The design patent is illustrated in Figures 3.1 and 3.2; a plant patent is shown in Figures 3.3 to 3.5; and a utility patent is shown in Figures 3.6 to 3.10.

Design and plant patents are special categories of patents that have been added as the technology and need for special types of patents became required. They make up a small portion of the U.S. patents granted. Over 400,000 design patents and over 10,000 plant patents have been issued. These are important to the industries they protect and are the basis of much litigation and many royalty battles.

The basis on which the patents are issued by the PTO is established by the laws passed by Congress and implemented by the rules of the PTO. These rules are established under procedures that allow the public to provide input and advice before they are put into effect. They may be suggested by the courts in their decisions or by the general public.

The courts play a major role in the patent system, as they are the interpreters of the laws concerning patents. Litigation between inventors is quite common — determining infringement, royalties payments, and interpretations of claims. These lawsuits are brought in federal court, because they involve U.S. patents. In years past, these lawsuits were brought forward in the federal district courts around the country, and it was common knowledge that certain courts were *pro*-patent, while others were *anti*-patent. Effort was made by the party bringing the lawsuit to select a court that would be favorable to its side. Court decisions were not uniform, and no assurance could be placed on the worth or value of a patent. As a result, industry's opinion of the value of a patent decreased. In 1984, a law was passed that altered the judicial system in the U.S. and established a Court of Appeals for the Federal Circuit (CAFC). This court was established to have jurisdiction over patent cases and is staffed by personnel who are skilled in the patent area. Now, all patent cases that are appealed from federal district courts automatically go to the CAFC (Figure 3.11). This court also serves as a possible recipient of an appeal from PTO actions. The appeal from a PTO decision may, on the discretion of the inventor, go to the U.S. District Court of the District of Columbia or directly to the CAFC. The CAFC is one step below the Supreme Court; and if a case is appealed from the CAFC, it goes directly to the Supreme Court. Since its establishment, decisions in patent cases have become more uniform, thus strengthening the protection that a

‖‖‖‖‖‖‖‖‖‖‖‖‖‖‖‖‖‖‖‖‖‖‖‖‖‖‖‖‖‖‖
US00D407853S

# United States Patent [19]

**Morgan**

[11] **Patent Number:** **Des. 407,853**

[45] **Date of Patent:** **∗∗∗Apr. 6, 1999**

[54] **SINGLE PAN COMPACT**

[75] Inventor: **Stephen Morgan**, Buckinghamshire, United Kingdom

[73] Assignee: **Maybelline Cosmetics Corp.**, New York, N.Y.

[ ∗ ] Notice: The term of this patent shall not extend beyond the expiration date of Pat. No. Des. 404,533.

[∗∗] Term: **14 Years**

[21] Appl. No.: **67,545**

[22] Filed: **Mar. 11, 1997**

[51] **LOC (6) Cl.** ............................................ **28-03**
[52] **U.S. Cl.** ............................................ **D28/78**
[58] **Field of Search** ..................... D28/76–84; D9/341, D9/418, 323; 132/293–307, 314, 315; 206/235, 581, 823, 37, 38; 220/DIG. 26

[56] **References Cited**

### U.S. PATENT DOCUMENTS

| | | | |
|---|---|---|---|
| D. 248,675 | 7/1978 | Gaudiche | D28/78 |
| D. 290,297 | 6/1987 | Schemmer | D28/78 |
| D. 302,745 | 8/1989 | Bakic | D28/83 |
| D. 303,582 | 9/1989 | Gavin | D28/78 |
| D. 364,485 | 11/1995 | Gavin | D28/78 |
| D. 365,892 | 1/1996 | Markham | D28/83 |
| D. 370,306 | 5/1996 | Stevens | D28/78 |
| 4,538,725 | 9/1985 | Glover et al. | 206/37 |
| 5,163,457 | 11/1992 | Lombardi, Jr. | 206/235 X |

### FOREIGN PATENT DOCUMENTS

| | | | |
|---|---|---|---|
| 589773 | 3/1994 | European Pat. Off. | 401/88 |
| 2161789 | 1/1986 | United Kingdom | 206/823 |

### OTHER PUBLICATIONS

Vogue Jul. 1983, p. 32—compacts indicated by arrows, Jan. 1986.
Modern Salon Jan. 1986, p. 30: compact at top of page.
Vogue Jan. 1986, p. 275—compacts on lower right.
Vogue Sep. 1992, p. 474: item No. 3 on lower left.
Hong Kong Enterprise Jun. 1994, p. 609—compact on left.
Glamour Sep. 1994, Revlon advertising insert with coupon: compact in foreground.
Chanel advertising leaflet, c. 1996—Double Teint Poudre compact.
1. Orange Plastic Co., Rahaway, N.J. 07065, Item 240 Mini S/L–undated.
2. Corona Plastics, Inc., Denville, NJ, #111–D–undated.
3. Jerhel Plastics, Inc., Bayonne, NJ #111–undated.
4. Package Works, Inc., #DK316–undated.
5. Shore Plastics, Inc., Freeport, NY 11520, #2300–undated.
6. E.B. Kingman Co., Leominster, MA 01453–undated.
7. Corona Plastics, Inc., Denville, NJ, #112–B–undated.

*Primary Examiner*—Ted Shooman
*Assistant Examiner*—C Tuttle
*Attorney, Agent, or Firm*—Baker & Botts, LLP

[57] **CLAIM**

The ornamental design for the single pan compact, as shown and described.

**DESCRIPTION**

FIG. 1 is a three-dimensional perspective view of the single pan compact showing my new design;
FIG. 2 is a front view thereof;
FIG. 3 is a top view thereof;
FIG. 4 is a side view thereof;
FIG. 5 is a back view thereof; and,
FIG. 6 is a bottom view thereof.

**1 Claim, 1 Drawing Sheet**

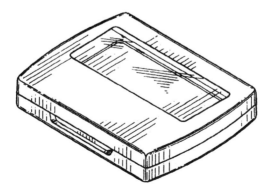

*Figure 3.1* Title page of a design patent, U.S. Des. 407,853.

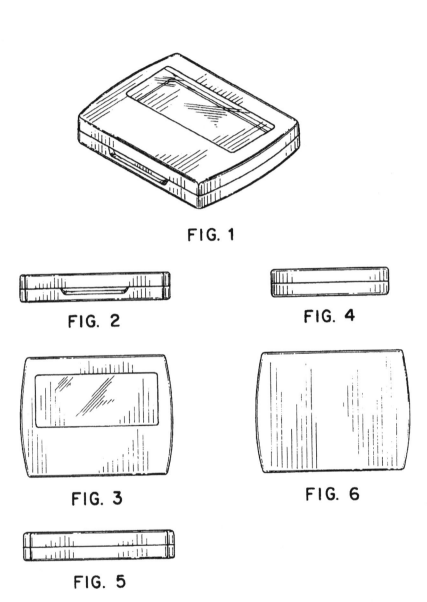

*Figure 3.2*    Design patent, Des. 407,853, sheet of drawings.

patent is intended to provide. The CAFC has provided a guide to the business community as to what it can expect from a patent grant.

One tenet of the U.S. Patent System was set soon after the system was founded: when two inventors claim the same invention, the one who was

US00PP10872P

# United States Patent [19]

Zaiger et al.

[11] Patent Number: **Plant 10,872**

[45] **Date of Patent:** **Apr. 27, 1999**

[54] PEACH TREE NAMED 'KLONDIKE WHITE'

[76] Inventors: **Chris Floyd Zaiger,** 929 Grimes Ave.; **Gary Neil Zaiger,** 1907 Elm Ave.; **Leith Marie Gardner,** 1207 Grimes Ave.; **Grant Gene Zaiger,** 4005 California Ave., all of Modesto, Calif. 95358

[21] Appl. No.: **08/982,135**

[22] Filed: **Dec. 1, 1997**

[51] Int. Cl.⁶ ............................................... A01H 5/00

[52] U.S. Cl. ............................................... Plt./196

[58] Field of Search ............................... Plt./42.1, 196

*Primary Examiner*—Howard J. Locker
*Assistant Examiner*—Anne Marie Grünberg

[57]                ABSTRACT

A new and distinct variety of peach tree (*Prunus persica*) which has the following unique combination of desirable features that are outstanding in a new variety. The following features of the tree and its fruit are characterized with the tree budded on nemaguard rootstock, grown on Hanford sandy loam soil with Storie Index rating 95, in USDA hardiness zone 9, near Modesto, Calif., and with standard commercial cultural fruit growing practices, such as, pruning, thinning, spraying, irrigation, fertilization, etc.:

1. Heavy and regular production of large size fruit.
2. Fruit with a mild, sweet, sub-acid flavor and excellent eating quality.
3. Fruit with firm white flesh, good handling and shipping quality.
4. Relatively uniform large-sized fruit throughout the tree.
5. Fruit with a high degree of attractive red skin color.
6. The fruit holding firm on the tree 7 to 10 days after maturity (shipping ripe).

1 Drawing Sheet

**1**

ORIGIN OF THE VARIETY

The present new and distinct variety of peach tree was developed by us in our experimental orchard located near Modesto, Calif. as a first generation cross between a selected seedling (field identification number 37G890) and May Crest Peach (U.S. Plant Pat. No. 4,064). The female parent (37G890) originated from a cross of Ruby Gold Nectarine (U.S. Plant Pat. No. 3,101) with a white peach of unknown parentage. A large number of these first generation seedlings growing on their own root were planted and maintained under close observation, during which time one such seedling evidenced desirable tree and fruit characteristics was selected for asexual propagation and commercialization.

ASEXUAL REPRODUCTION OF THE VARIETY

Asexual reproduction of the new and distinct variety of peach tree was by budding on nemaguard rootstock, as performed by us in our experimental orchard located near Modesto, Calif., and shows that all characteristics of the tree and its fruit run true to the original tree and are established and transmitted through succeeding asexual propagations.

SUMMARY OF THE VARIETY

The new and distinct variety of peach tree which is of large size, vigorous, upright growth; and a productive and regular bearer of large size, white flesh, freestone fruit. The firm flesh of the fruit is mild, sweet, sub-acid, with excellent flavor and eating quality, further characterized to its novelty is its relatively uniform size fruit throughout the tree, its good handling and shipping quality, having attractive red skin color and, in comparison to Giant Babcock Peach (U.S. Plant Pat. No. 1,353), the fruit of the new variety has firmer flesh, greater handling and shipping quality and is approximately 2 weeks earlier in maturity.

PHOTOGRAPH OF THE VARIETY

The accompanying color photographic illustration shows typical specimens of the foliage and fruit of the present new peach variety. The illustration shows the upper and lower

**2**

surface of the leaves, an exterior and sectional view of a fruit divided in its suture plane to show flesh color, pit cavity and the stone remaining in place. The photographic illustration was taken shortly after being picked (shipping ripe) and the colors are as nearly true as is reasonably possible in a color representation of this type.

DESCRIPTION OF THE VARIETY

The following is a detailed botanical description of the new variety of peach tree, its flowers, foliage and fruit, as based on observations of specimens grown near Modesto, Calif., with color terminology (except those in common terms) in accordance with Reinhold Color Atlas by A. Kornerup and J. H. Wanscher.

Tree:

*Size.*—Large — tree height controlled by selective pruning to 12 to 14 feet in height and 12 feet to 14 feet in width primarily for economical harvesting.

*Vigor.*—Vigorous — grows 6 to 8 feet in height the first growing season. During first dormant season the tree is pruned to 4 to 5 feet in height and primary scaffolds are selected that will carry a heavy crop load. Height and width of tree is controlled by pruning, primarily in the dormant season.

*Form.*—Usually pruned to vase shape.

*Productivity.*—Productive — fruit set is 1½ or more times the amount desirable for normal tree crop load.

*Bearer.*—Regular — each year fruit must be thinned and spaced to reduce crop load to the number that will make desirable market size fruit. Number of fruit desired per tree varies with age of tree, tree spacing, cultural practices, soil type and climatic conditions.

*Density.*—Medium dense — shoot growth and leaves restrict sunlight and air movement in the center of the tree. The tree is pruned to a vase shape allowing more sunlight and air movement which enhances skin color and fruit with higher Brix.

*Figure 3.3*   Title page of a plant patent, U.S. Plant 10, 872.

first to invent shall receive the patent. Or, as it is commonly expressed, the system is based on a "first-to-invent" concept. Other countries do not follow this system, but have established a "first-to-file" system; that is, the first

Plant 10,872

**3**

*Growth.*—Upright.
*Hardiness.*—Winter chilling requirement is approximately 850 hours below 45° F. Hardiness tested only in USDA Hardiness Zone 9.
Trunk:
  *Size.*—Medium.
  *Texture.*—Medium shaggy.
  *Color.*—Mouse gray to soot brown (5-E-5) to (5-F-4).
Branches:
  *Size.*—Medium.
  *Texture.*—Smooth to medium rough, varies with age of growth.
  *Lenticels.*—Numerous. Medium size.
  *Color.*—Light brown to brown (5-D-6) to (5-E-4), varies with age of growth.
Leaves:
  *Size.*—Large. Average length 6". Average width 1½".
  *Form.*—Lanceolate. Pointed.
  *Margin.*—Crenate.
  *Thickness.*—Medium.
  *Surface.*—Slightly crinkled, new leaves somewhat smoother.
  *Petiole.*—Medium length — ½". Medium thickness.
  *Glands.*—Globose. Number varies from 1 to 4. Average number 2. Medium size. Located on upper portion of petiole and lower portion of leaf blade.
  *Color.*—Upper surface — jade green to green (27-E-6) to (27-E-8). Lower surface — dull green to light green (27-E-4) to (27-E-5).
Flower buds:
  *Size.*—Large.
  *Length.*—Medium.
  *Form.*—Plump.
  *Pubescence.*—Pubescent.
Flowers:
  *Size.*—Large, showy. 1¼" to 1⅜" in diameter.
  *Pollen.*—Present, self-fertile.
  *Blooming period.*—Date of first bloom: Mar. 4, 1996. Date of last bloom: Mar. 9, 1996. Varies slightly with climatic conditions.
  *Color.*—Pink to light pink (10-A-3) to (10-A-2). Pink fades to light pink with age of flowers.
Fruit:
  *Maturity when described.*—Firm ripe.
  *Date of first picking.*—Jun. 25, 1996.
  *Date of last picking.*—Jul. 2, 1996. Varies slightly with climatic conditions.
  *Size.*—Large. Average diameter axially 2⅞". Average transversely in suture plane to 3". Average weight 210 grams. Weight varies from 204 to 220 grams.
  *Form.*—Nearly globose.
  *Suture.*—Shallow, extends from base to apex.
  *Ventral surface.*—Rounded.
  *Apex.*—Usually rounded, varies from rounded to slight apical tip.
  *Base.*—Retuse.
  *Cavity.*—Rounded to slightly elongated in suture plane. Average depth ½". Average breadth ¾".
Flesh:
  *Ripens.*—Evenly.
  *Texture.*—Firm.
  *Fibers.*—Small, tender.
  *Aroma.*—Slight.
  *Amygdalin.*—Undetected.
  *Eating quality.*—Excellent.
  *Flavor.*—Excellent, mild, sweet, sub-acid.

**4**

*Juice.*—Moderate amount — mild, sub-acid, enhances flavor. Average Brix 11.5°. Brix varies with amount of crop load on the tree and climatic conditions — warm temperatures and sunlight increases Brix.
  *Color.*—White to yellowish white (1-A-1) to (1-A-2). Pit cavity — greenish white to grayish yellow (1-B-3) to (1-B-4). Very slight bleeding of light red from pit cavity into flesh.
Stem:
  *Size.*—Medium. *Average length* ½". Average diameter ⅛". Enlarged at point of fruit attachment.
  *Color.*—Yellowish green to grayish green (1-B-5) to (1-C-5).
Skin:
  *Thickness.*—Medium.
  *Texture.*—Medium, tenacious to the flesh.
  *Down.*—Moderate, short in length.
  *Tendency to crack.*—None.
  *Color.*—Yellowish white to pastel yellow ground color (1-A-3) to (2-A-4). Overspread with crayfish red to garnet red (9-D-8) to (10-D-8). Fruit with more exposure to sunlight have a greater degree of enhanced red color.
Stone:
  *Type.*—Freestone.
  *Size.*—Average length 1½". Average width 1⅛". Average thickness ¾".
  *Form.*—Obovoid.
  *Base.*—Usually straight, varies from straight to rounded.
  *Apex.*—Usually acuminate, varies from acuminate to rounded.
  *Surface.*—Furrowed toward apex, pitted toward base, both pits and furrows relatively deep. Pits vary from round to elongated. Ridges relatively wide at surface of stone.
  *Sides.*—Unequal. One half of the stone is slightly larger in size, extending farther from the suture line, with slightly deeper furrows and pit cavities.
  *Tendency to split.*—None.
  *Color.*—Light brown to brown (6-D-4) to (6-D-6).
Use: Dessert. Market, local and long distance.
Keeping quality: Good — flesh texture remains firm after 10 days in cold storage.
Shipping quality: Good — minimal bruising of flesh or skin scarring during picking and shipping trials.

The present new variety of peach tree, its flowers, foliage and fruit herein described may vary in slight detail due to climate, soil conditions and cultural practices under which the variety may be grown. The present description is that of the variety grown under the ecological conditions prevailing near Modesto, Calif.
We claim:
1. A new and distinct variety of peach tree, substantially as illustrated and described, characterized by its large size, vigorous, upright growth; and a productive and regular bearer of large, firm, white flesh, freestone fruit with good handling and shipping quality; the fruit is further characterized by having a mild, sweet, sub-acid taste with excellent flavor and eating quality, having a high degree of attractive red skin color and, in comparison to the Giant Babcock Peach (U.S. Plant Pat. No. 1,353), the new variety has firmer flesh with greater shipping quality and is approximately 2 weeks earlier in maturity.

\* \* \* \* \*

*Figure 3.4*   Plant patent, Plant 10,872, specifications.

applicant to file is awarded the patent, and ownership of the invention is left to court action. There is a movement in the U.S. to alter the system and go to a first-to-file system.

**U.S. Patent**            **Apr. 27, 1999**            **Plant 10,872**

*Figure 3.5*   Plant patent, Plant 10,872. A photograph (provided in color in the application) is used to show the product.

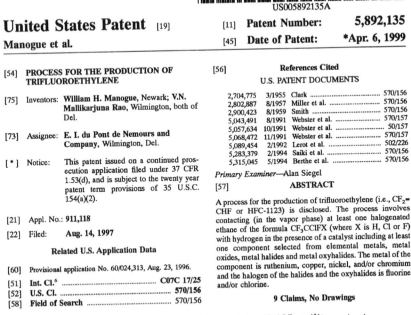

Figure 3.6   Title page of U.S. Patent, 5,892,135, utility patent.

Another tenet that is dominant in the U.S. system is the criteria for a patent. It is stated in the law that for an invention to be patentable:

- The invention must be truly new.
- The invention must not be obvious in light of what has been known before.
- The invention must be useful.

Each will be discussed in more detail in Chapter 5, "The Patentable Invention." The U.S. Patent and Trademark Office, located in Arlington, VA, maintains for the public at that location files of all U.S. patents, files of patents of many foreign countries, and a technical library. It grants patents and trademarks and publishes other documents to assist the promotion and use of patents and trademarks.

The PTO is staffed with patent examiners to evaluate each application. These examiners are required to have a degree in science or technology and are assigned to an area that is appropriate to their college training. Many of the examiners are also attorneys, and many have advanced scientific degrees. For many years, the examining corps has been divided into three major areas: chemical, mechanical, and electrical. Each of these areas was subdivided into smaller, more specialized groups. These three divisions have now been elim-

5,892,135

**1**

## PROCESS FOR THE PRODUCTION OF TRIFLUOROETHYLENE

This application claims the priority benefit of U.S. Provisional application Ser. No. 60/024,313, filed Aug. 23, 1996.

### FIELD OF THE INVENTION

This invention relates to a process for the production of trifluoroethylene, and more particularly, to a catalytic process for the preparation of trifluoroethylene from saturated halogenated hydrocarbons.

### BACKGROUND

Trifluoroethylene (i.e., $CF_2=CHF$ or HFC-1123) is a useful monomer for the preparation of fluorocarbon polymers.

HFC-1123 can be produced from $CCl_2FCClF_2$ (CFC-113) by reaction with hydrogen in the presence of a catalyst comprising palladium and at least one metal selected from gold, tellurium, antimony, bismuth and arsenic (U.S. Pat. No. 5,283,379). HFC-1123 can also be prepared from $CF_2=CClF$ (CFC-1113) by reaction with hydrogen in the presence of a catalyst comprising palladium or platinum on a magnesium oxide carrier (U.S. Pat. No. 5,089,454). A product of over-hydrogenolysis from both CFC-113 and CFC-1113 is $CH_2FCHF_2$ (HFC-143). There is continuing interest in developing efficient processes for the manufacture of HFC-1123.

### SUMMARY OF THE INVENTION

This invention provides a process for the production of trifluoroethylene (i.e., $CF_2=CHF$ or HFC-1123). The process comprises contacting in the vapor phase at least one halogenated ethane of the formula $CF_3CClFX$ where X is selected from the group consisting of H, Cl and F, with hydrogen in the presence of a catalyst comprising at least one component selected from the group consisting of elemental metals, metal oxides, metal halides and metal oxyhalides; wherein the metal of said component is selected from ruthenium, copper, nickel, chromium and mixtures thereof and the halogen of said halides and said oxyhalides is selected from fluorine, chlorine and mixtures thereof.

### DETAILED DISCUSSION

The present invention provides a process for the manufacture of trifluoroethylene by contacting at least one halogenated ethane selected from the group consisting of 2-chloro-1,1,1,2-tetrafluoroethane (i.e., $CHClFCF_3$ or HCFC-124), 2,2-dichloro-1,1,1,2-tetrafluoroethane (i.e., $CCl_2FCF_3$ or CFC-114a) and chloropentafluoroethane (i.e., $CClF_2CF_3$ or CFC-115) with hydrogen. The contact is in the vapor phase and in the presence of selected hydrogenation catalysts. Preferred processes include those which use the halogenated ethane HCFC-124 and/or the halogenated ethane CFC-114a as starting material. The more preferred starting material is HCFC-124. It is noted that CFC-114 (i.e., $CClF_2CClF_2$) and HCFC-124a (i.e., $CClF_2CHF_2$) can also be employed as starting materials under conditions where they isomerize (to CFC-114a and HCFC-124, respectively). Catalysts which facilitate such isomerization include catalysts comprising aluminum halides (e.g., $AlF_3$) or aluminum oxyhalides.

The present invention involves the use of advantageously catalytic components employing ruthenium, copper, nickel and/or chromium. Suitable components include elemental

**2**

metals such as ruthenium and ruthenium-copper mixtures; halides such as CuF, CuCl, CuClF, $NiF_2$, $NiCl_2$, NiClF, $CrF_3$, $CrCl_3$, $CrCl_2F$ and $CrClF_2$; oxides such as CuO, NiO, and $Cr_2O_3$; and oxyhalides such as copper oxyfluoride and chromium oxyfluoride. Oxyhalides may be produced by conventional procedures such as, for example, halogenation of metal oxides.

The catalysts of this invention may contain other components, some of which are considered to improve the activity and/or longevity of the catalyst composition. Preferred catalysts include catalysts which are promoted with compounds of molybdenum, vanadium, tungsten, silver, iron, potassium, cesium, rubidium, barium or combinations thereof. Also of note are chromium-containing catalysts which further contain zinc and/or aluminum or which comprise copper chromite.

The catalyst may be supported or unsupported. Supports such as metal fluorides, alumina and titania may be advantageously used. Particularly preferred are supports of fluorides of metals of Group IIB, especially calcium. A preferred catalyst consists essentially of copper, nickel and chromium oxides (each of said oxides being preferably present in equimolar quantities) preferably promoted with potassium salt, on calcium fluoride.

An especially preferred catalyst contains proportionally about 1.0 mole CuO, about 0.2 to 1.0 mole NiO, about 1 to 1.2 moles $Cr_2O_3$ on about 1.3 to 2.7 moles $CaF_2$, promoted with about 1 to 20 weight %, based on the total catalyst weight, of an alkali metal selected from K, Cs, and Rb (preferably K). When K is the promoter, the preferred amount is from about 2 to 15 weight % of the total catalyst.

This catalyst can be prepared by coprecipitating, from an aqueous medium, salts of copper, nickel and chromium (and optionally aluminum and zinc), with and preferably on calcium fluoride; washing, heating and drying the precipitate. An alkali metal compound (e.g., KOH, KF or $K_2CO_3$) is then deposited on the dried precipitate, followed by calcination to convert the copper, nickel and chromium to the respective oxides. Any soluble copper, nickel and chromium compound may be used, but the fluorides, chlorides and nitrates are preferred, with the nitrates being especially preferred. Alternatively, promoters such as KOH, KF and $K_2CO_3$ may be added prior to co-precipitation.

Another group of catalysts which may be used for the conversion of $CF_3CClFX$ contains proportionally about 1.0 mole CuO, about 0.2 to 1.0 mole NiO, about 1 to 1.2 moles $Cr_2O_3$, about 0.4 to 1.0 mole $MoO_3$, and about 0.8 to 4.0 mole $CaF_2$, optionally promoted with at least one compound from the group consisting of $MgF_2$, $MnF_2$ and $BaF_2$. Pd or $WO_3$ may also be present.

Another preferred group of catalysts are those comprising ruthenium. Of note are supported ruthenium catalysts. Examples include ruthenium supported on aluminum fluoride, chromium fluorides, rare earth fluorides, or divalent metal fluorides.

The catalyst may be granulated, pressed into pellets, or shaped into other desirable forms. The catalyst may contain additives such as binders and lubricants to help insure the physical integrity of the catalyst during granulating or shaping the catalyst into the desired form. Suitable additives include carbon and graphite. When binders and/or lubricants are added to the catalyst, they normally comprise about 0.1 to 5 weight percent of the weight of the catalyst.

The catalyst may be activated prior to use by treatment with hydrogen, air, or oxygen at elevated temperatures. After use for a period of time in the process of this invention, the

*Figure 3.7* U.S. Patent 5,892,135, first page of specifications.

5,892,135

**3**

activity of the catalyst may decrease. When this occurs, the catalyst may be reactivated by treating it with hydrogen, air or oxygen, at elevated temperature in the absence of organic materials.

The molar ratio of hydrogen to $CF_3CClFX$ typically ranges from about 0.5:1 to about 30:1, and is preferably within the range of about 1:1 to about 24:1.

The process of the present invention is suitably conducted at a temperature in the range of from about 200° C. to 500° C., preferably from about 325° C. to about 425° C. The contact time of reactants with the catalyst bed is typically from about 0.1 seconds to about 2.0 minutes, and is preferably from about 10 seconds to 90 seconds.

Atmospheric or superatmospheric pressures may suitably be employed.

The reaction products may be separated by conventional techniques such as distillation. In accordance with this invention, when $CF_3CHClF$ is contacted with catalyst, the reaction temperature, pressure and contact time may be controlled such that the major (selectivity greater than 50%) reaction product is trifluoroethylene. Other products which can be isolated and are useful are tetrafluoroethane (i.e., $CH_2FCF_3$ or HFC-134a), a refrigerant, and vinylidene fluoride (i.e., $CH_2=CF_2$ or HFC-1132a), a monomer for making fluorinated polymers.

The process of this invention can be carried out readily in the vapor phase using well known chemical engineering practice.

The reaction zone and its associated feed lines, effluent lines and associated units should be constructed of materials resistant to hydrogen fluoride and hydrogen chloride. Typical materials of construction, well-known to the fluorination art, include stainless steels, in particular of the austenitic type, the well-known high nickel alloys, such as Monel® nickel-copper alloys, Hastelloy® nickel-based alloys and, Inconel® nickel-chromium alloys, and copper-clad steel. Silicon carbide is also suitable for reactor fabrication.

Without further elaboration, it is believed that one skilled in the art can, using the description herein, utilize the present invention to its fullest extent. The following preferred specific embodiments are to be construed as illustrative, and not as constraining the remainder of the disclosure in any way whatsoever.

### EXAMPLES

Catalyst Preparation for Examples 1–5

Aqueous calcium nitrate (2.7 moles) is mixed with aqueous potassium fluoride (5.4 moles), heated and stirred briefly at 100° C. to form a slurry of $CaF_2$. To this slurry is added copper nitrate (1 mole), nickel nitrate (1 mole) and chromium nitrate (1 mole) as solids. The slurry is stirred at 70° to 80° C. until the salts, other than $CaF_2$, dissolve. This is followed by adding 0.1 mole of aqueous potassium hydroxide over 1 hour and boiling the mixture briefly. The slurry is cooled to 40° to 50° C. and filtered. The solid is washed exhaustively to reduce the potassium content to an undetectable level. After drying, potassium hydroxide is added as a solution in quantities sufficient to provide a catalyst containing 9 weight % potassium. After drying again, the catalyst is calcined at 600° C. for 8 to 16 hours, then granulated and screened to 1 to 2 mm particles. The catalyst is mixed with 1 to 5 wt % "Sterotex" powdered lubricant (registered trademark of Capital City Products Co., Columbus Ohio, division of Stokely-Van Camp, for its edible hydrogenated vegetable oil) to give ⅛"x⅛" (3.2 mmx3.2

**4**

mm) cylindrical pellets from a Stokes tablet machine. This catalyst is used as described in Examples 1–5.

General Procedure for Product Analysis for Examples 1–4

The products leaving the reactor were analyzed on line using a gas chromatograph. The column consisted of a 20′ (6.1 m)x⅛"(3.2 mm) s/s tube containing Krytox™ perfluorinated polyether on an inert support. Helium was used as the carrier gas. The product analyses are reported in mole %.

| Legend | |
|---|---|
| 124 is $CHClFCF_3$ | 143a is $CH_3CF_3$ |
| 1132a is $CH_2=CF_2$ | 1123 is $CHF=CF_2$ |
| 134a is $CH_2FCF_3$ | 1122 is $CHCl=CF_2$ |

CT is the contact time in minutes at reaction conditions, based on catalyst volume

Conv. is starting material converted

Selectivities are based upon the mole fraction of starting material converted

### EXAMPLE 1

A 15" (38.1 cm)x¼" (0.64 cm) O.D. Inconel™ 600 nickel alloy U-tube reactor was charged with catalyst (8.3 g, 6 mL) pellets prepared substantially in accordance with the Catalyst Preparation above. The catalyst was treated with 50 sccm (8.3x10⁻⁷ m³/s) air at 525° C. for 1.0 hours and at 400° C. for 0.33 hours, followed by purging with 150 sccm (2.5x10⁻⁶ m³/s) nitrogen at 400° C. for 5 hours, and $H_2$ were contacted with the catalyst at 300° to 425° C., 0 psig (101 kPa) and with a molar ratio of $H_2$:$CHClFCF_3$ of 1:1. Results of the reaction for 23 hours are shown in the

TABLE 1

| Run | Temp. | CT | % Conv. | Selectivities, % | | | |
|---|---|---|---|---|---|---|---|
| No. | °C. | min. | 124 | 1132a | 1123 | 134a | 1122 |
| 1 | 300 | 1.27 | 34 | 4.2 | 67 | 24 | 3.4 |
| 2 | 325 | 1.21 | 53 | 4.3 | 74 | 15 | 3.9 |
| 3 | 350 | 1.17 | 74 | 6.4 | 78 | 10 | 2.5 |
| 4 | 375 | 1.12 | 83 | 7.4 | 82 | 6 | 1.2 |
| 5 | 400 | 1.08 | 91 | 6.9 | 83 | 5 | 0.5 |
| 6 | 425 | 1.04 | 84 | 7.9 | 82 | 5 | 0.5 |

### EXAMPLE 2

A 15" (38.1 cm)x⅜" (0.95 cm) O.D. Inconel™ 600 nickel alloy U-tube reactor was charged with the catalyst (21.7 g, 18 mL) pellets prepared substantially in accordance with the Catalyst Preparation above. Following prior use of the catalyst, the catalyst was treated with air at 300° to 400° C. for 10 hours, followed by purging with nitrogen at 400° C. for 6 hours. The catalyst was then treated with mixture of 50 sccm (8.3x10⁻⁷ m³/s) hydrogen and 100 sccm (1.7x10⁻⁶ m³/s) nitrogen at 400° to 450° C. for 1.5 hours and finally purging with nitrogen while cooling the catalyst bed to 300° C. $CHClFCF_3$ and $H_2$ were contacted with the catalyst at 300° to 340° C., 0 psig (101 kPa) and with a molar ratio of $H_2$:$CHClFCF_3$ of 4:1. Results of the reaction are shown in the Table 2.

*Figure 3.8*    U.S. Patent 5,892,135, second page of specifications and examples.

5,892,135

## 5

### TABLE 2

| Run | Temp. | CT | % Conv. | Selectivities, % | | |
|-----|-------|-----|---------|------|------|------|
| No. | °C. | min. | 124 | 1132a | 1123 | 134a |
| 1 | 300 | 0.5 | 30 | 4.6 | 68 | 19 |
| 2 | 320 | 0.5 | 45 | 7.5 | 72 | 15 |
| 3 | 340 | 0.5 | 65 | 7.6 | 77 | 10 |

### EXAMPLE 3

The same reactor and catalyst were used as used in Example 2. CHClFCF$_3$ and H$_2$ were contacted with the catalyst at 330° to 350° C., 30 psig (308 kPa) and with a molar ratio of H$_2$:CHClFCF$_3$ of 4:1. Results of the reaction are shown in the Table 3.

### TABLE 3

| Run | Temp. | CT | % Conv. | Selectivities, % | | |
|-----|-------|-----|---------|------|------|------|
| No. | °C. | min. | 124 | 1132a | 1123 | 134a |
| 1 | 330 | 0.5 | 45 | 6.5 | 71 | 16 |
| 2 | 330 | 1.0 | 58 | 5.0 | 73 | 15 |
| 3 | 340 | 0.5 | 56 | 5.7 | 72 | 15 |
| 4 | 340 | 1.0 | 67 | 4.8 | 74 | 14 |
| 5 | 350 | 0.5 | 65 | 5.8 | 74 | 13 |
| 6 | 350 | 1.0 | 73 | 4.8 | 75 | 13 |

### EXAMPLE 4

The same reactor and catalyst were used as used in Example 3. The catalyst was purged with nitrogen at 200° C. for 45 minutes. The catalyst was then treated with 50 sccm (8.3×10$^{-7}$ m$^3$/s) air at 300° C. for 3.5 hours, at 350° C. for 5 hours, at 400° C. for 6 hours, followed by purging with nitrogen at 400° C., for 6 hours. CHClFCF$_3$ and H$_2$ were contacted with the catalyst at 340° to 380° C., 300 psig (2169 kPa) and with a molar ratio of H$_2$:CHClFCF$_3$ of 24:1. The results of the reaction are shown in the Table 4. The potassium-promoted catalyst sample used in this example had been in use for a total of over 600 hours (not including regeneration time) after completion of the runs in Table 4.

### TABLE 4

| Run | Temp. | CT | % Conv | Selectivities, % | | | | | |
|-----|-------|-----|--------|------|------|------|------|------|------|
| No. | °C. | min. | 124 | 1132a | 1123 | 143a | 134a | 1122 |
| 1 | 340 | 0.47 | 77 | 2.6 | 61 | 2.4 | 26 | 1.1 |
| 2 | 340 | 0.94 | 87 | 2.5 | 54 | 2.6 | 31 | 0.9 |
| 3 | 350 | 0.23 | 59 | 4.5 | 59 | 2.5 | 26 | 1.1 |
| 4 | 360 | 0.45 | 78 | 4.4 | 58 | 2.5 | 26 | 1.2 |
| 5 | 360 | 0.23 | 59 | 5.0 | 60 | 2.6 | 24 | 1.1 |
| 6 | 350 | 0.24 | 46 | 4.8 | 60 | 2.5 | 26 | 1.1 |
| 7 | 350 | 0.92 | 75 | 3.6 | 59 | 2.3 | 26 | 1.1 |
| 8 | 355 | 0.46 | 56 | 4.1 | 63 | 2.3 | 22 | 1.1 |
| 9 | 365 | 0.45 | 53 | 3.6 | 70 | 1.5 | 16 | 1.1 |
| 10 | 380 | 0.45 | 45 | 2.7 | 79 | 0.0 | 9 | 1.1 |

Experimental Procedure for Examples 5–7

The reaction of HCFC-124 and H$_2$ was carried out using a multi-port vapor phase reactor. A nickel cylinder was bored out to hold eight individual catalyst tubes. Each tube, when connected with the rest of the reactor system, had its own nitrogen purge supply. Each of the catalyst tubes could be charged with catalyst prior to sealing the nickel cylinder and

## 6

wrapping it in insulation. Thermocouples were placed on the surface of the cylinder and in an interior location within the cylinder to monitor temperature.

HCFC-124 and H$_2$ were fed to the reactor system using a Tylan® 280SA mass flow meter. The mass flow meter was calibrated using a J&W Scientific ADM1000 Flow Meter.

For each example, the catalyst was used as received without pre-treatment for a nitrogen purge. After 2 hours of run time, the products leaving the reactor were analyzed on-line using a Hewlett-Packard® 5880 gas chromatograph. The column used was a 20'x⅛" stainless steel tube containing Krytox™ perfluorinated polyether on an inert support. Helium was used as the carrier gas. Product analyses are listed in mole %. Selectivity is based upon the mole % of HCFC-124 converted.

### EXAMPLE 5

A 3" (7.6 cm)x¼" (0.64 cm) O.D. Inconel™ 600 nickel alloy tube was charged with 1.11 grams (0.9 cc) of catalyst prepared substantially in accordance with the Catalyst Preparation above and then crushed to about 20–30 mesh. Hydrogen was fed at 2 sccm and HCFC- 124 was fed at 1 sccm (corresponding to a contact time of about 0.3 min). The reactor temperature was set at 400° C. After 2 hours, the conversion of HCFC-124 was 39% and the selectivity to HFC-1123 was 85%.

### EXAMPLE 6

Catalyst used was 2% ruthenium on EuF$_3$. The catalyst was prepared by impregnating EuF$_3$ with a solution of RuCl$_3$, and reducing with H$_2$ at 300° C. The particle size was about 20–30 mesh. 1.27 grams (0.86 cc) of catalyst was charged into a tube as described in Example 5. Results of the reaction are shown in Table 5. H$_2$ and HCFC-124 feed rates were varied. The reactor was set at 350° C.

### TABLE 5

| Temp (°C.) | 124 Flow (sccm) | H$_2$ Flow (sccm) | Conversion HCFC-124 | Selectivity 1123 |
|------|------|------|------|------|
| 350 | 10.0 | 5.0 | 14.2% | 61.7% |
| 350 | 5.0 | 5.0 | 21.4% | 55.5% |
| 350 | 5.0 | 2.5 | 16.2% | 53.6% |

### EXAMPLE 7

Example 6 was substantially repeated except that the catalyst used was Ru-CrF$_3$ prepared by collapsing a (NH$_3$)$_6$RuCrF$_6$ composition (see U.S. patent application Ser. No. 60/007,734 and PCT International Publication No. WO 7/19751). 0.74 grams (0.86 cc) of catalyst was charged into a tube as described in Example 5. H$_2$ and HCFC-124 feed rates, as well as the reactor temperatures, were varied. Results of the reaction are shown in Table 6.

### TABLE 6

| Temp (°C.) | 124 Flow (sccm) | H2 Flow (sccm) | Conversion HCFC-124 | Selectivity 1123 |
|------|------|------|------|------|
| 300 | 5 | 20 | 56.4% | 21.6% |
| 300 | 5 | 5 | 55.7% | 23.2% |
| 300 | 5 | 2.5 | 45.1% | 32.1% |
| 300 | 10 | 5 | 35.8% | 35.0% |
| 325 | 10 | 5 | 27.2% | 74.9% |
| 300 | 10 | 5 | 28.5% | 88.3% |

*Figure 3.9* U.S. Patent 5,892,135, additional examples and tables.

5,892,135

**7**

We claim:

1. A process for the production of trifluoroethylene, comprising:

contacting in the vapor phase at least one halogenated ethane of the formula $CF_3CClFX$ where X is selected from the group consisting of H, Cl and F, with hydrogen in the presence of a catalyst comprising at least one component selected from the group consisting of elemental metals, metal oxides, metal halides and metal oxyhalides; wherein the metal of said component is selected from ruthenium, copper, nickel, chromium and mixtures thereof and the halogen of said halides and said oxyhalides is selected from fluorine, chlorine and mixtures thereof.

2. The process of claim 1 wherein halogenated ethane selected from the group consisting of $CF_3CHClF$, $CF_3CCl_2F$ and mixtures thereof is used as starting material.

3. The process of claim 2 wherein the starting material is $CF_3CHClF$.

4. The process of claim 3 wherein the catalyst contains proportionally about 1.0 mole CuO, about 0.2 to 1.0 mole

**8**

NiO, about 1 to 1.2 moles $Cr_2O_3$ on about 1.3 to 2.7 moles $CaF_2$, and is promoted with about 1 to 20 weight %, based on the total catalyst weight, of an alkali metal selected from K, Cs, and Rb.

5. The process of claim 4 wherein the catalyst is promoted with from about 2 to 15 weight % K, based on the total catalyst weight.

6. The process of claim 3 wherein the catalyst contains proportionally about 1.0 mole CuO, about 0.2 to 1.0 mole NiO, about 1 to 1.2 moles $Cr_2O_3$, about 0.4 to 1.0 mole $MoO_3$, and about 0.8 to 4.0 mole $CaF_2$.

7. The process of claim 6 wherein the catalyst is promoted with at least one compound selected from the group consisting of $MgF_2$, $MnF_2$, and $BaF_2$.

8. The process of claim 3 wherein the catalyst comprises ruthenium.

9. The process of claim 3 wherein the catalyst consisting essentially of copper, nickel and chromium oxides, promoted with potassium salt, on calcium fluoride.

\*   \*   \*   \*   \*

*Figure 3.10*    U.S. Patent 5,892,135, allowed claims.

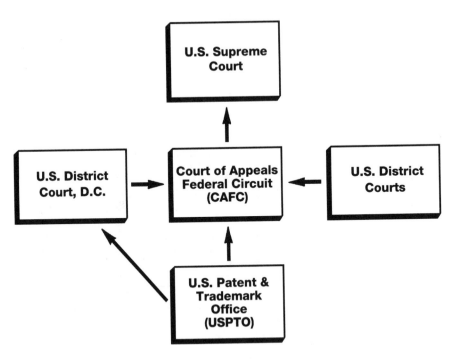

*Figure 3.11*    Federal court system of appeals for patent cases.

inated, and six technology areas will be established to handle the operations. This has been done because technology has changed, and science areas are becoming less definable. The six areas, or technology centers, are:

Sector 1: Biotechnology, Organic Chemistry, and Designs
Sector 2: Chemical and Material Engineering
Sector 3: Transportation, Construction, and Agriculture
Sector 4: Mechanical Engineering, Manufacturing, and Products
Sector 5: Communications and Information Processing
Sector 6: Physics, Optics, System Components, and Electrical Engineering

Examiners are barred from obtaining a patent of their own during their tenure at the PTO and, if they leave governmental service, they are barred for a time from prosecuting an application in their old examining group.

The PTO functions as a large bureaucratic operation, with departments and functions designed to handle the administrative needs of the organization. The organizational chart of the PTO is shown in Figure 3.12. As one can see from the organization, many functions relate to the relationship between the PTO and Congress, as well as foreign patent organizations. The PTO representatives function as U.S. representatives to official meetings with foreign patent organizations and individual patent offices of foreign countries.

PTO operations are centralized in the Crystal City area of Arlington, VA, and the offices are distributed in 16 buildings. Some of the locations are distant from each other. In an attempt to improve operations, a plan is under way to consolidate the entire function into one group of buildings devoted exclusively to the Patent and Trademark Office. Money has been appropriated to plan for such a complex. Site selection is under way, and bids are being considered to build the buildings and move the entire operation into the new complex by 2005. When the location is selected, the modification of plans for the construction can begin.

At the same time that this centralization project was in progress, another plan was also under consideration by Congress to convert the PTO from a branch of the Commerce Department into an independent agency of the U.S. Government, similar to the Federal Communications Commission, the Postal Service, etc. This approach would also, according to its backers, improve the efficiency of operations. It would enable the PTO to function as a truly "pay as you go" business operation by eliminating the federal rules that restrict the employment regulations, the purchasing of equipment, and allow the management to better manage the operations. The proposal was acted upon by the U.S. Congress in 1999. Congress passed this legislation on the last day of the 1999 session. All sections of the law will be effective in 2000. On March 29, 2000, the PTO became a Performance Based Organization, with the relaxing of many federal rules.

Since PTO records cover a period of over 200 years and all of the patent files are available to the public, they must be protected. The dissemination of information is another important function of the PTO and is increasing each year. In patent application matters, *all* business is conducted in a written format and in a confidential manner between the applicant and the examiner. The contents of the correspondence are not available to the public until the

## U.S. Patent and Trademark Organizational Chart

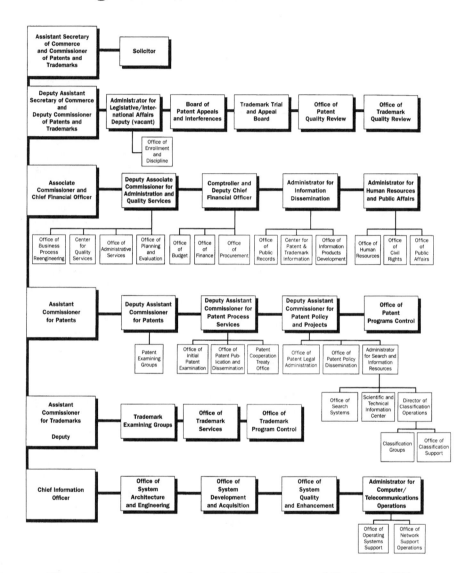

*Figure 3.12*  Organization chart of the U.S. Patent and Trademark Office.

patent issues, at which time the file of the patent becomes public information. In the event that an application does not become a patent, the application must be retained but is not available to the public. The mass of papers generated is gigantic, and all must be stored in a safe and secure manner.

It has been mentioned that over 6 million utility patents have been issued since the patent system originated. Who are the inventors? Are they all

**Table 3.1**  Applications Filed and Patents Issued

|  | 1995 | 1996 | 1997 | 1998 |
|---|---|---|---|---|
| Applications filed | 236,679 | 206,276 | 237,045 | 240,090 |
| Patents issued | 114,241 | 116,657 | 122,977 | 154,647 |
| Patents issued to U.S. residents | 64,562 | 66,498 | 69,294 | 85,843 |
|  | (56.51%) | (57.00%) | (56.35%) | (55.50%) |
| Patents issued to foreign residents | 49,679 | 50,159 | 53,683 | 68,804 |
|  | (43.39%) | (43.00%) | (43.65%) | (44.50%) |

**Table 3.2**  Organizations Receiving Most U.S. Patents in 1998

| 1998 Rank | Organization | Number of Patents |
|---|---|---|
| 1 | **International Business Machines** | **2657** |
| 2 | Canon Kabushike Kaisha | 1928 |
| 3 | NEC Corporation | 1627 |
| 4 | **Motorola Inc.** | **1406** |
| 5 | Sony Corporation | 1316 |
| 6 | Samsung Electronics Co. Ltd. | 1304 |
| 7 | Fujitsu Limited | 1189 |
| 8 | Toshiba Corporation | 1170 |
| 9 | **Eastman Kodak Company** | **1124** |
| 10 | Hitachi, Ltd. | 1094 |

Americans? Are they primarily inventors working in their basements or garages, or are they the large industrial companies with huge research facilities? About 150,000 patents are issued each year, many to foreign applicants. The ratio of foreign applicants to U.S. applicants has been climbing through the years; Table 3.1 provides data for the past few years.

Which industrial corporations obtain the most patents? Table 3.2 summarizes recent results.

IBM has topped the list of corporations awarded the most U.S. patents since 1994; and it is typical for only a few — in this case three — U.S. corporations to be in the Top 10 list. If the agencies of the U.S. Government were combined, they would have received 1017 patents in 1998 and would place number 12 on the list in Table 3.2.

Who in the U.S. obtains patents? The inventors come from every state and all of the territories. In 1998, the top five states that led the list in numbers of patents were:

1. California
2. New York
3. Texas
4. Illinois
5. New Jersey

Records have not been kept on the number of individual inventors; but by looking at the copies of the *Official Gazette,* the weekly publication of the

PTO that lists all patents issued that week, one can see that many of the patents are issued to individuals and are not assigned to a corporation. The PTO has recently created The Office of Independent Inventor Programs, a new office that will be directed to meet the needs of independent inventors. A recent survey of applications filed showed that 15% fell into the category of independent inventors. This office has established a Web Site on the Internet, and will provide additional information about the PTO and its procedures through brochures, flyers, and materials for agent and attorney selection.

Attempts are under way to simplify the paperwork. Plans for computer-generated applications and documents have been discussed, but many problems remain to be solved before that can become a reality. The major ones are the security of the application, the ability to make copies, permanence, and assurance as to the integrity of the records. Progress is being made in this direction. Experimental programs on electronic application filing have been tested and are now available for trademark applications. Standardized forms for various patent operations can now be obtained via the Internet. More services will follow quickly.

## chapter four

# The Canadian patent system

In 1987, as the leaders of the world's patent community talked of change, Canada's Parliament moved ahead with a nearly total revision of its traditional patent system. Until then, the Canadian patent system operated in principle and in practice in a manner very similar to that of the United States. Pending applications were maintained in secrecy; there was no publication until (and unless) the patent was granted; and the patent term was 17 years from the date of grant. Perhaps the most noteworthy feature common to the U.S. system and the traditional Canadian system — in sharp contrast to most of the world's patent systems — was the "first-to-invent" concept.

Among the basic changes incorporated into the new Canadian Patent Act were:

- The Canadian patent system is now a "first-to-file" system.
- The term of a patent is 20 years from the date of application in Canada.
- Pending applications are open to public inspection before issue.
- Examination may be deferred for a period of up to 7 years from the filing date.
- Maintenance fees are required, starting on the second anniversary of the Canadian filing date.

The transition from the old system to the new required an interim period during which patents were issued under both systems. Applications filed before October 1, 1989, were processed under the old system and will have a term of 17 years from the issue date. Applications filed after that date are being processed under the new system, and the issued patent will have a term of 20 years from the filing date.

Prior to the passage of the new law, there was a substantial backlog of pending applications. The backlog was increased considerably by a rush to file additional cases under the old law. In the interim, patents were issued under both systems. Those issuing under the old law were numbered in the usual sequential manner, while patents issuing under the new law have followed a numbering system that started with 2,000,000, so that it will be easy to determine the laws that apply and the expiration date of the patent.

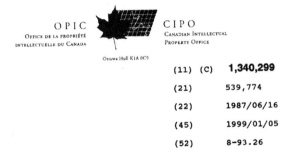

O P I C
OFFICE DE LA PROPRIÉTÉ
INTELLECTUELLE DU CANADA

C I P O
CANADIAN INTELLECTUAL
PROPERTY OFFICE

Ottawa-Hull K1A 0C9

| | | |
|---|---|---|
| (11) | (C) | **1,340,299** |
| (21) | | 539,774 |
| (22) | | 1987/06/16 |
| (45) | | 1999/01/05 |
| (52) | | 8-93.26 |

(51) Int.Cl.[6]   D06M 13/12; D06M 13/52

(19)(CA) **CANADIAN PATENT** (12)

(54) Process for Making Individualized Crosslinked Fibers and
Fibers Thereof

(72) Schoggen, Howard L.  , U.S.A.
Cook, Jeffrey T.  , U.S.A.
Bourbon, Robert M.  , U.S.A.

(73) Buckeye Cellulose Corporation (The) , U.S.A.

(30) (US) U.S.A.   879,678 1986/06/27

(57) 17 Claims

NO DRAWING

▮✦▮ Industrie   Industry
Canada   Canada

OPIC - CIPO 191

Canadä

---

*Figure 4.1*   Title page of Canadian Patent 1,340,299, issued January 5, 1999, under
the old rules.

Some 10 years have passed since the inception of the new law and rules,
but the backlog of cases under the old laws is still being acted upon.
Figure 4.1 shows the title page of a patent (1,340,299) that was issued under
the old rules on January 5, 1999. Figure 4.2 shows a patent (2,058,483) that
was issued the same day under the new rules.

OPIC
Office de la propriété
intellectuelle du Canada

CIPO
Canadian Intellectual
Property Office

(12)(19)(CA) **Brevet-Patent**

(11)(21)(C) **2,058,483**
(22)   1991/12/27
(43)   1992/07/01
(45)   1999/01/05

(72) Norman, John A. T., US
(72) Dyer, Paul N., US
(73) Air Products and Chemicals, Inc., US
(51) Int.Cl.$^6$ C23C 16/18
(30) 1990/12/31 (07/636316) US
(54) **PROCEDE POUR LE DEPOT D'UNE PELLICULE DE CUIVRE A PARTIR D'UN COMPLEXE VOLATIL**
(54) **PROCESS FOR THE CHEMICAL VAPOR DEPOSITION OF COPPER**

$$Cu^{+1} \ (R^1 - \overset{\overset{\displaystyle O}{\|}}{C} - \overset{\overset{\displaystyle R^2}{|}}{C} - \overset{\overset{\displaystyle O}{\|}}{C} - R^3)^{-1} \bullet L$$

(57) Divulgation d'un processus pour déposer sélectivement des pellicules de cuivre sur des parties conductrices de l'électricité en métal ou en une autre matière de surfaces de substrat en mettant en contact le substrat, à une température variant entre 110 et 190 °C, avec un complexe de cuivre organométallique volatil, en phase gazeuse, représenté par la formule structurale : (voir formule ci-haut), dans laquelle $R^1$ et $R^3$ sont indépendamment l'un de l'autre un perfluoroalkyle en $C_1$-$C_8$, $R^2$ est un atome d'hydrogène ou un perfluoroalkyle en $C_1$-$C_8$, et L est un monoxyde de carbone, un isonitrile ou un ligand hydrocarbure non saturé contenant au moins une non-saturation non aromatique.

(57) A process is provided for selectively depositing copper films on metallic or other electrically conducting portions of substrate surfaces by contacting the substrate at a temperature from 110° to 190°C with a volatile organometallic copper complex, in the gas phase, represented by the structural formula: (see above formula) where $R^1$ and $R^3$ are each independently $C_1$-$C_8$ perfluoroalkyl, $R^2$ is H or $C_1$-$C_8$ perfluoroalkyl and L is carbon monoxide, an isonitrile, or an unsaturated hydrocarbon ligand containing at least one non-aromatic unsaturation.

Industrie Canada   Industry Canada

*Figure 4.2*   Title page of Canadian Patent 2,058,483, issued January 5, 1999, under the new rules.

The general rules of patentability — except where the new "first-to-file" concept applies — remain similar to those in the U.S. Other differences will be discussed later in this book. The basic procedures for preparing, filing, and processing patent applications remain very similar to those of the U.S.

A major exception is that, since Canada is a bilingual country, patent applications can be filed in either English or French. There is no provision for opposition; however, the validity of a patent can be attacked in court.

The present Canadian laws retain the requirement that a patent will be issued only to the inventor or inventors. A patent cannot be issued to one who derives knowledge of the invention from another. Under the first-to-file rules, when two parties file patent applications claiming the same invention, the application with the earliest actual Canadian filing date (or priority date) will be citable as a reference against the other application.

Canada has established a maintenance fee system for patent applications and patents. Maintenance fees begin on the second anniversary of the Canadian filing date and continue to the 19th anniversary. In the U.S., maintenance fees are due at the 3½-, 7½-, and 11½-year anniversaries of the date of issuance to keep the patent in force. Failure to pay a maintenance fee results in the loss of patent rights.

Canada had participated in the PCT (Patent Cooperation Treaty) organization since its inception but, due to legal restrictions, was unable to fully utilize the system. These problems were eliminated, and Canada became a full member of the PCT system in early 1990.

Under present law, a patent application is "laid open for public inspection" within 18 months of the earlier of its actual Canadian filing date or its priority date. At that time, copies of the application are made available to the public. The weekly publication of the Canadian Patent Office, the *Canadian Patent Office Record*, lists the patents issued during the week, as well as the applications laid open to the public. Figures 4.3, 4.4, and 4.5 show typical pages. The *Canadian Patent Office Record* shows titles of new patents but does not include abstracts or claims explaining the patents or applications — a contrast to the U.S. Patent Office weekly publication *Official Gazette.*

The "laying open an application" for public inspection allows a patent owner to sue one who has infringed the claims of a laid open patent application for reasonable damages (or other remedies) from the date that a patent issued. The differences between reasonable compensation and regular damages are left to the determination of the court.

Because the Canadian application is "laid open for inspection" within a definite time period, the secrecy of an application is maintained only until it is published. If an application is withdrawn, a timely request to the Canadian Patent Office can retain the application in secret.

The current law provides for a deferred examination. An applicant in Canada can delay the examination of the application for up to 7 years from the date of filing. In contrast, in the U.S. all applications are examined upon receipt. Provision has also been made to allow interested third parties to request examination. A reason for deferring the examination in Canada is to allow the applicant to concentrate, for example, on the prosecution of the corresponding U.S. patent application. Once the U.S. patent issues, the Canadian applicant can incorporate the U.S. claim amendments and hopefully gain expedited allowance of the Canadian application.

## CANADIAN PATENTS ISSUED
## January 05, 1999
## BREVETS CANADIENS DÉLIVRÉS
## le 05 janvier 1999

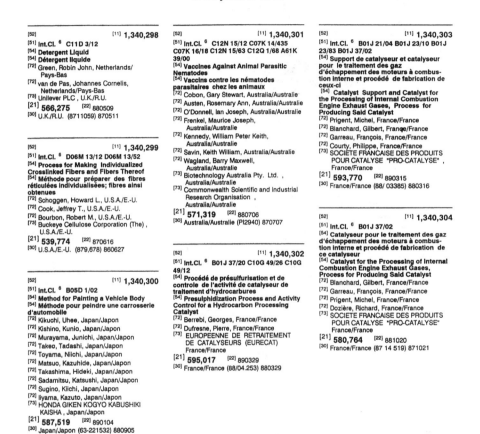

[52]     [11] **1,340,298**
[51] Int.Cl. 6 C11D 3/12
[54] Detergent Liquid
[54] Détergent liquide
[72] Green, Robin John, Netherlands/
Pays-Bas
[72] van de Pas, Johannes Cornelis,
Netherlands/Pays-Bas
[73] Unilever PLC , U.K./R.U.
[21] **566,275**   [22] 880509
[30] U.K./R.U. (8711059) 870511

[52]     [11] **1,340,299**
[51] Int.Cl. 6 D06M 13/12 D06M 13/52
[54] Process for Making Individualized
Crosslinked Fibers and Fibers Thereof
[54] Méthode pour préparer des fibres
réticulées individualisées; fibres ainsi
obtenues
[72] Schoggen, Howard L., U.S.A./E.-U.
[72] Cook, Jeffrey T., U.S.A./E.-U.
[72] Bourbon, Robert M., U.S.A./E.-U.
[73] Buckeye Cellulose Corporation (The) ,
U.S.A./E.-U.
[21] **539,774**   [22] 870616
[30] U.S.A./E.-U. (879,678) 860627

[52]     [11] **1,340,300**
[51] Int.Cl. 6 B05D 1/02
[54] Method for Painting a Vehicle Body
[54] Méthode pour peindre une carrosserie
d'automobile
[72] Kikuchi, Uhee, Japan/Japon
[72] Kishino, Kunio, Japan/Japon
[72] Murayama, Junichi, Japan/Japon
[72] Takeo, Tadashi, Japan/Japon
[72] Toyama, Niichi, Japan/Japon
[72] Matsuo, Kazuhide, Japan/Japon
[72] Takashima, Hideki, Japan/Japon
[72] Sadamitsu, Katsushi, Japan/Japon
[72] Sugino, Kiichi, Japan/Japon
[72] Iiyama, Kazuto, Japan/Japon
[73] HONDA GIKEN KOGYO KABUSHIKI
KAISHA , Japan/Japon
[21] **587,519**   [22] 890104
[30] Japan/Japon (63-221532) 880905

[52]     [11] **1,340,301**
[51] Int.Cl. 6 C12N 15/12 C07K 14/435
C07K 16/18 C12N 15/63 C12Q 1/68 A61K
39/00
[54] Vaccines Against Animal Parasitic
Nematodes
[54] Vaccins contre les nématodes
parasitaires chez les animaux
[72] Cobon, Gary Stewart, Australia/Australie
[72] Austen, Rosemary Ann, Australia/Australie
[72] O'Donnell, Ian Joseph, Australia/Australie
[72] Frenkel, Maurice Joseph,
Australia/Australie
[72] Kennedy, William Peter Keith,
Australia/Australie
[72] Savin, Keith William, Australia/Australie
[72] Wagland, Barry Maxwell,
Australia/Australie
[73] Biotechnology Australia Pty. Ltd. ,
Australia/Australie
[73] Commonwealth Scientific and Industrial
Research Organisation ,
Australia/Australie
[21] **571,319**   [22] 880706
[30] Australia/Australie (PI2940) 870707

[52]     [11] **1,340,302**
[51] Int.Cl. 6 B01J 37/20 C10G 49/26 C10G
49/12
[54] Procédé de présulfurisation et de
contrôle de l'activité de catalyseur de
traitement d'hydrocarbures
[54] Presulphidization Process and Activity
Control for a Hydrocarbon Processing
Catalyst
[72] Berrebi, Georges, France/France
[72] Dufresne, Pierre, France/France
[73] EUROPEENNE DE RETRAITEMENT
DE CATALYSEURS (EURECAT)
France/France
[21] **595,017**   [22] 890329
[30] France/France (88/04.253) 880329

[52]     [11] **1,340,303**
[51] Int.Cl. 6 B01J 21/04 B01J 23/10 B01J
23/83 B01J 37/02
[54] Support de catalyseur et catalyseur
pour le traitement des gaz
d'échappement des moteurs à combustion interne et procédé de fabrication de
ceux-ci
[54] Catalyst Support and Catalyst for
the Processing of Internal Combustion
Engine Exhaust Gases, Process for
Producing Said Catalyst
[72] Prigent, Michel, France/France
[72] Blanchard, Gilbert, France/France
[72] Garreau, François, France/France
[73] SOCIETE FRANCAISE DES PRODUITS
POUR CATALYSE "PRO-CATALYSE" ,
France/France
[21] **593,770**   [22] 890315
[30] France/France (88/ 03385) 880316

[52]     [11] **1,340,304**
[51] Int.Cl. 6 B01J 37/02
[54] Catalyseur pour le traitement des gaz
d'échappement des moteurs à combustion interne et procédé de fabrication de
ce catalyseur
[54] Catalyst for the Processing of Internal
Combustion Engine Exhaust Gases,
Process for Producing Said Catalyst
[72] Blanchard, Gilbert, France/France
[72] Garreau, François, France/France
[72] Prigent, Michel, France/France
[72] Dozière, Richard, France/France
[73] SOCIETE FRANCAISE DES PRODUITS
POUR CATALYSE "PRO-CATALYSE"
France/France
[21] **580,764**   [22] 881020
[30] France/France (87 14 519) 871021

*Figure 4.3*   Page from the *Canadian Patent Office Record,* the weekly publication of the Canadian Patent Office, showing patents issued under the old rules.

Unlike the U.S., Canada allows an inventor or his/her assignee to file a patent application. If the applicant is the assignee, an original or notarized copy of the assignment of the Canadian right to the invention must be filed with the Canadian Patent Office. In Canada, an assignment must include a witness to the inventor's signature.

The subject matter acceptable for patenting in Canada is essentially the same as in the U.S. Several notable exceptions occur. Canada will allow only certain classes of life-forms to be patented. Although the U.S. has pioneered this area, Canada is not following the broad scope of U.S. patents. Also, inventions relating to methods or processes for treating living humans or

**CANADIAN PATENTS ISSUED**
**January 05, 1999**

[52]　　　　　　　　　[11] **2,057,949**
[51] Int.Cl. 6　G09F 9/30
[54] DISPLAY DEVICE
[54] DISPOSITIF D'AFFICHAGE
[72] Killinger, Erich, Germany (Federal Republic of)/Allemagne (République Fédérale d')
[73] Dambach-Werke GmbH , Germany (Federal Republic of)/Allemagne (République Fédérale d')
[21] **2,057,949**　　[22] 911218
[30] Germany (Federal Republic of)/ Allemagne (République Fédérale d') (P 40 40 567.2-32) 901219

[52]　　　　　　　　　[11] **2,058,154**
[51] Int.Cl. 6　A61B 17/435
[54] SET OF INSTRUMENTS FOR THE UTERINAL EMBRYO TRANSFER AND INTRA-UTERINE INSEMINATION
[54] ENSEMBLE D'INSTRUMENTS POUR LE TRANSFERT D'UN EMBRYON DANS L'UTERUS ET L'INSEMINATION INTRA-UTERINE
[72] Schinkel, Otto, Germany (Federal Republic of)/Allemagne (République Fédérale d')
[72] Hervath, Josef, Germany (Federal Republic of)/Allemagne (République Fédérale d')
[73] Labotect-Labor-Technik Göttingen GmbH , Germany (Federal Republic of)/Allemagne (République Fédérale d')
[21] **2,058,154**　　[22] 911219
[30] Germany (Federal Republic of)/ Allemagne (République Fédérale d') (G 91 07 790.3) 910625

[52]　　　　　　　　　[11] **2,058,483**
[51] Int.Cl. 6　C23C 16/18
[54] PROCESS FOR THE CHEMICAL VAPOR DEPOSITION OF COPPER
[54] PROCEDE POUR LE DEPOT D'UNE PELLICULE DE CUIVRE A PARTIR D'UN COMPLEXE VOLATIL
[72] Norman, John A. T., U.S.A./E.-U.
[72] Dyer, Paul N., U.S.A./E.-U.
[73] Air Products and Chemicals, Inc. U.S.A./E.-U.
[21] **2,058,483**　　[22] 911227
[30] U.S.A./E.-U. (07/636316) 901231

[52]　　　　　　　　　[11] **2,058,893**
[51] Int.Cl. 6　A61M 1/00
[54] TRACHEAL SUCTION CATHETER
[54] CATHETER POUR SUCCION TRACHEALE
[72] Russo, Ronald D., U.S.A./E.-U.
[73] Russo, Ronald D. , U.S.A./E.-U.
[21] **2,058,893**　　[22] 920110

[52]　　　　　　　　　[11] **2,059,712**
[51] Int.Cl. 6　C23C 2/02
[54] GALVANIZED HIGH-STRENGTH STEEL SHEET HAVING LOW YIELD RATIO AND METHOD OF PRODUCING THE SAME
[54] FEUILLE D'ACIER GALVANISE HAUTE-MENT RESISTANTE ET METHODE DE FABRICATION
[72] Masui, Susumu, Japan/Japon
[72] Sakata, Kei, Japan/Japon
[72] Togashi, Fusao, Japan/Japon
[73] KAWASAKI STEEL CORPORATION , Japan/Japon
[21] **2,059,712**　　[22] 920120
[30] Japan/Japon (44580/1991) 910121

[52]　　　　　　　　　[11] **2,060,888**
[51] Int.Cl. 6　D04H 1/54
[54] POLYOLEFIN STRETCH NON-WOVEN FABRIC AND METHOD OF MAKING IT
[54] TEXTILE NON TISSE EXTENSIBLE DU POLYOLEFINE, ET METHODE POUR SA FABRICATION
[72] Nakajima, Takayoshi, Japan/Japon
[72] Yokota, Seiji, Japan/Japon
[73] CHISSO CORPORATION , Japan/Japon
[73] Uni-Charm Co., Ltd. , Japan/Japon
[21] **2,060,888**　　[86] 910501
[30] Japan/Japon (2-111848) 900501

[52]　　　　　　　　　[11] **2,061,876**
[51] Int.Cl. 6　C25B 9/00 C25B 1/26 C02F 1/76
[54] MEMBRANELESS CHLORINE GAS GENERATING APPARATUS
[54] APPAREIL POUR LA FABRICATION DE GAZ CHLORE SANS MEMBRANE
[72] Keller, Robert D., Jr., U.S.A./E.-U.
[73] Keller, Robert D., Jr. , U.S.A./E.-U.
[21] **2,061,876**　　[22] 920226
[30] U.S.A./E.-U. (07/741,631) 910807
[30] U.S.A./E.-U. (07/662,922) 910301

[52]　　　　　　　　　[11] **2,063 '8**
[51] Int.Cl. 6　F41A 25/12
[54] FIREARM PARTICULARLY A HAND GUN
[54] ARME A FEU, DE TYPE PISTOLET
[72] Möller, Tilo, Germany (Federal Republic of)/Allemagne (République Fédérale d')
[72] Brandl, Rudolf, Germany (Federal Republic of)/Allemagne (République Fédérale d')
[72] Weidle, Helmut, Germany (Federal Republic of)/Allemagne (République Fédérale d')
[72] Krieger, Hubert, Germany (Federal Republic of)/Allemagne (République Fédérale d')
[73] Heckler & Koch GmbH , Germany (Federal Republic of)/Allemagne (République Fédérale d')
[21] **2,063,178**　　[22] 920317
[30] Germany (Federal Republic of)/ Allemagne (République Fédérale d') (P 41 09 777.7) 910325

[52]　　　　　　　　　[11] **2,063,415**
[51] Int.Cl. 6　C08F 297/02
[54] COPOLYMERES TRISEQUENCES ACRYLIQUES, LEUR PREPARATION, ET LEUR APPLICATION A LA FABRICATION D'ARTICLES ELASTOMERES
[54] ACRYLIC TRISEQUENCED COPOLYMERS; PROCESS FOR PREPAR-ING THE SAME AND THEIR USE FOR MANUFACTURING ELASTOMERIC AR-TICLES
[72] Varshnéy, Sunil K., Belgique/Belgium
[72] Fayt, Roger, Belgique/Belgium
[72] Teyssie, Philippe, Belgique/Belgium
[72] Hautekeer, Jean-Paul, Belgique/Belgium
[73] ELF ATOCHEM S.A. , France/France
[21] **2,063,415**　　[86] 900705
[30] France/France (89/09268) 890710
[30] France/France (89/15581) 891127

[52]　　　　　　　　　[11] **2,063,516**
[51] Int.Cl. 6　H04Q 3/42 H04L 12/56
[54] A SELF-ROUTING SWITCH BROAD-CAST METHOD
[54] METHODE DE DIFFUSION PAR COMMUTATEUR D'AUTOACHEMINEMENT
[72] Uriu, Shiro, Japan/Japon
[72] Yoshimura, Shuji, Japan/Japon
[72] Kakuma, Satoshi, Japan/Japon
[73] FUJITSI LIMITED , Japan/Japon
[21] **2,063,516**　　[22] 920319
[30] Japan/Japon (03-056921) 910320

*Figure 4.4*　Page from the *Canadian Patent Office Record*, showing patents issued under the new rules.

animals by surgery or therapy have been held to be unpatentable. They have ruled that a method of medical treatment that lies in the field of professional surgery cannot be considered a "process" or "art" within the meaning of the patent act. Articles of manufacture or apparati designed for use in the treatment of humans or animals, or diagnostic methods, can be patented.

## CANADIAN LAID-OPEN APPLICATIONS
### November 22, 1998 - November 28, 1998

## DEMANDES CANADIENNES MISES À LA DISPONIBILITÉ DU PUBLIC
### 22 novembre 1998 - 28 novembre 1998

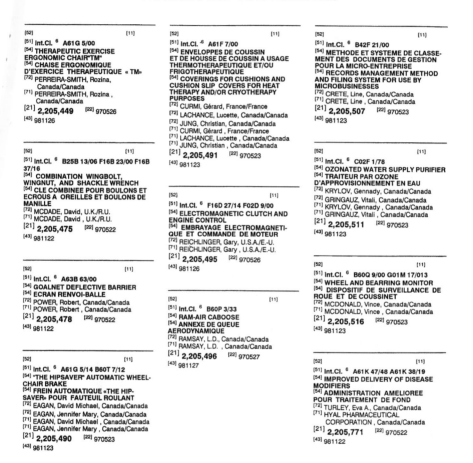

[52]                                                          [11]
[51] Int.Cl. [6] A61G 5/00
[54] THERAPEUTIC EXERCISE
ERGONOMIC CHAIR"TM"
[54] CHAISE ERGONOMIQUE
D'EXERCICE THERAPEUTIQUE « TM»
[72] PERREIRA-SMITH, Rozina,
Canada/Canada
[71] PERREIRA-SMITH, Rozina ,
Canada/Canada
[21] **2,205,449**    [22] 970526
[43] 981126

[52]                                                          [11]
[51] Int.Cl. [6] B25B 13/06 F16B 23/00 F16B
37/16
[54] COMBINATION WINGBOLT,
WINGNUT, AND SHACKLE WRENCH
[54] CLE COMBINEE POUR BOULONS ET
ECROUS A OREILLES ET BOULONS DE
MANILLE
[72] MCDADE, David, U.K./R.U.
[71] MCDADE, David , U.K./R.U.
[21] **2,205,475**    [22] 970522
[43] 981122

[52]                                                          [11]
[51] Int.Cl. [6] A63B 63/00
[54] GOALNET DEFLECTIVE BARRIER
[54] ECRAN RENVOI-BALLE
[72] POWER, Robert, Canada/Canada
[71] POWER, Robert , Canada/Canada
[21] **2,205,478**    [22] 970522
[43] 981122

[52]                                                          [11]
[51] Int.Cl. [6] A61G 5/14 B60T 7/12
[54] "THE HIPSAVER" AUTOMATIC WHEEL-
CHAIR BRAKE
[54] FREIN AUTOMATIQUE «THE HIP-
SAVER» POUR FAUTEUIL ROULANT
[72] EAGAN, David Michael, Canada/Canada
[72] EAGAN, Jennifer Mary, Canada/Canada
[71] EAGAN, David Michael , Canada/Canada
[71] EAGAN, Jennifer Mary , Canada/Canada
[21] **2,205,490**    [22] 970523
[43] 981123

[52]                                                          [11]
[51] Int.Cl. [6] A61F 7/00
[54] ENVELOPPES DE COUSSIN
ET DE HOUSSE DE COUSSIN A USAGE
THERMOTHERAPEUTIQUE ET/OU
FRIGOTHERAPEUTIQUE
[54] COVERINGS FOR CUSHIONS AND
CUSHION SLIP COVERS FOR HEAT
THERAPY AND/OR CRYOTHERAPY
PURPOSES
[72] CURMI, Gérard, France/France
[72] LACHANCE, Lucette, Canada/Canada
[72] JUNG, Christian, Canada/Canada
[71] CURMI, Gérard , France/France
[71] LACHANCE, Lucette , Canada/Canada
[71] JUNG, Christian , Canada/Canada
[21] **2,205,491**    [22] 970523
[43] 981123

[52]                                                          [11]
[51] Int.Cl. [6] F16D 27/14 F02D 9/00
[54] ELECTROMAGNETIC CLUTCH AND
ENGINE CONTROL
[54] EMBRAYAGE ELECTROMAGNETI-
QUE ET COMMANDE DE MOTEUR
[72] REICHLINGER, Gary, U.S.A./E.-U.
[71] REICHLINGER, Gary , U.S.A./E.-U.
[21] **2,205,495**    [22] 970526
[43] 981126

[52]                                                          [11]
[51] Int.Cl. [6] B60P 3/33
[54] RAM-AIR CABOOSE
[54] ANNEXE DE QUEUE
AERODYNAMIQUE
[72] RAMSAY, L.D., Canada/Canada
[71] RAMSAY, L.D. , Canada/Canada
[21] **2,205,496**    [22] 970527
[43] 981127

[52]                                                          [11]
[51] Int.Cl. [6] B42F 21/00
[54] METHODE ET SYSTEME DE CLASSE-
MENT DES DOCUMENTS DE GESTION
POUR LA MICRO-ENTREPRISE
[54] RECORDS MANAGEMENT METHOD
AND FILING SYSTEM FOR USE BY
MICROBUSINESSES
[72] CRETE, Line, Canada/Canada
[71] CRETE, Line , Canada/Canada
[21] **2,205,507**    [22] 970523
[43] 981123

[52]                                                          [11]
[51] Int.Cl. [6] C02F 1/78
[54] OZONATED WATER SUPPLY PURIFIER
[54] TRAITEUR PAR OZONE
D'APPROVISIONNEMENT EN EAU
[72] KRYLOV, Gennady, Canada/Canada
[72] GRINGAUZ, Vitali, Canada/Canada
[71] KRYLOV, Gennady , Canada/Canada
[71] GRINGAUZ, Vitali , Canada/Canada
[21] **2,205,511**    [22] 970523
[43] 981123

[52]                                                          [11]
[51] Int.Cl. [6] B60Q 9/00 G01M 17/013
[54] WHEEL AND BEARRING MONITOR
[54] DISPOSITIF DE SURVEILLANCE DE
ROUE ET DE COUSSINET
[72] MCDONALD, Vince, Canada/Canada
[71] MCDONALD, Vince , Canada/Canada
[21] **2,205,516**    [22] 970523
[43] 981123

[52]                                                          [11]
[51] Int.Cl. [6] A61K 47/48 A61K 38/19
[54] IMPROVED DELIVERY OF DISEASE
MODIFIERS
[54] ADMINISTRATION AMELIOREE
POUR TRAITEMENT DE FOND
[72] TURLEY, Eva A., Canada/Canada
[71] HYAL PHARMACEUTICAL
CORPORATION , Canada/Canada
[21] **2,205,771**    [22] 970522
[43] 981122

*Figure 4.5*    Page from the *Canadian Patent Office Record*, showing the "laid open" applications.

Canada does not issue "Design Patents," as the U.S. does. Instead, protection for aesthetic features applied to functional objects can be protected under the Industrial Design Act. An application for an industrial design must be filed within 1 year of publication in Canada. The registration will provide exclusive rights for a period of 5 years and may be maintained for an additional 5 years upon payment of a fee.

Other specific aspects of the Canadian law that differ from U.S. practice will be discussed in later chapters — when U.S. law is discussed.

In early 1999, Canada placed its issued patents on the World Wide Web (see Chapter 11 for address), allowing for complete searching and all images online. Copies of the patents can also be obtained through their office:

Canadian Intellectual Property Office
50 Victoria Street
Place du Portage 1
Hull, Quebec, Canada K1A 0C9

*chapter five*

# The patentable invention

"Whoever invents or discovers any *new and useful pro-cess, machine, manufacture, or composition of matter*, or any new and useful improvement thereof, may obtain a patent therefor, subject to the conditions and requirements of this title."

35 USC 101 (emphasis added)

Invention, whether patentable or not, can be viewed as the product of a problem-solving process. Typically, an inventor recognizes a problem or need. The creative solution to the problem is his/her invention. However, regardless of how well the invention solves the problem or satisfies the need, or how creative the invention is, for it to be a *patentable* invention, it must meet the requirements set forth in the patent laws.

In particular, the invention must be directed to appropriate subject matter, and it must exhibit three essential qualities: novelty, utility, and non-obviousness. Section 101 of the patent statutes specifies two of the essential qualities of a patentable invention: *new and useful*. A third essential quality, *non-obviousness*, is specified in another section. In addition, Section 101 defines the categories of subject matter that are suitable for patent protection:

Patentable Subject Matter

To be patentable, an invention must be:

- A process
- A machine
- A manufactured article, or
- A composition of matter.

If the subject matter of the invention does not fall within one of these four statutory classes, it is not suitable for patent protection.

The term "process" means process or method and may be thought of as one or more steps performed on a material, a composition, or an article to produce a change in its nature or characteristics to produce a useful product. Examples of processes include a process for making a chemical compound, a process for treating leather, a process for fabricating metal parts, or a method to improve commerce.

The term "machine" includes various mechanical devices and may be thought of as a group of elements or parts that interact to produce an intended effect or result. Examples of machines include a dishwasher, a carburetor, a lawn mower, or a washing machine.

The word "manufacture" means an article of manufacture or a manufactured article and includes practically anything made by humans. Examples of manufactured articles include such things as a toothbrush, a table, a bench, a golf ball, a transgenic animal, etc.

A "composition of matter" may be either a chemical compound or a mixture of ingredients, such as a formulation. Examples of patentable mixtures include such formulations as a toothpaste, a shampoo, a cleaning solution, an adhesive, etc.

Often, a close look at the creative efforts that led to one patentable invention will disclose related inventions that may fall within one or more of the other statutory classes of invention. For example, the invention of a new composition of matter, such as a chemical compound, may also involve the invention of a process for making the compound. Since the compound is probably intended for a specific use, there may, at the same time, be patentable inventions directed to the use of the compound in a process or as a material for an article of manufacture.

## Novelty requirement

A patentable invention must be novel. In general, this means that it must not have been known or used by someone in the U.S., or published or patented anywhere in the world before being invented by the person applying for a patent. Furthermore, to be novel, it must not have been published or patented anywhere in the world, or in use or on sale in the U.S. more than 1 year prior to the date of application for a patent. This is a way of describing that the invention must be new. This description defines what is meant by "prior art." In the U.S., if the invention is known in the prior art, then the invention is not novel, and thus not patentable. In Canada, knowledge available to the public in the 1-year period prior to the actual filing date of the patent application is *not* considered part of the prior art.

## Utility requirement

The utility requirement simply means that the invention must have some useful purpose. The examiner will, upon reading the patent application, determine that the application states a use for the invention, or the invention has a useful purpose. If the examiner does not feel he can answer those questions in a positive manner, he has the power to reject the application as inoperative, or on the basis of lack of utility. The applicant can refute the examiner's opinion by the presentation of a working model, testimony of experts, or test results. Rejection on utility is not very common.

## Non-obviousness requirement

The third requirement for patentability — and the one most often debated — is that the invention be non-obvious. The patent statutes require that the *differences* between the *subject matter* sought to be patented and the *prior art* (i.e., public knowledge, prior publications, patents, etc.) are such that the *subject matter as a whole* would *not have been obvious* at the time the invention was made *to a person having ordinary skill in the art* to which the subject matter pertains. Obviousness is a very subjective factor, and the issue of obviousness is the major source of disagreement between patent applicants and patent examiners.

In 1966, in a landmark patent case (Graham v. John Deere), the Supreme Court dealt with the question of obviousness. In its decision, the court identified the factors to be considered:

- The scope and content of the prior art
- The differences between the prior art and the claims at issue
- The level of ordinary skill in the art

Thus, if the differences between what was previously known (prior art) and the invention claimed are such that the invention would have been obvious to one of ordinary skill in the art, the invention is considered obvious and not patentable.

The Supreme Court further identified three secondary considerations that can be utilized in a determination of obviousness of an invention. They are:

- Commercial success
- Long felt, but unresolved needs
- Failure of others in the art

Thus, a showing of commercial success, a recognition that the invention fills a long-recognized but unresolved need, and/or the failure of others who tried to solve the problem can contribute to a finding of non-obviousness.

This decision has been reviewed by many courts and has established the obviousness for the patent community. However, it has not stopped the arguments between inventors and examiners. No — it is still being litigated!

## chapter six

# The evolution of a patent

The evolution of a patent begins with invention. What follows is a process that involves a merging of technical and legal efforts, guided at key points by business considerations. The process is outlined in an idealized flowchart in Figure 6.1. The chart is designed to illustrate a smooth path from conception to the granting of a patent, with no problems along the way. Each block highlights a significant event in the path, but it must be understood that, in the real world, each step could have multiple arrows representing possible pitfalls and alternate outcomes.

For convenience, the process can be considered in three stages:

1. The pre-patent application stage: from conception to decision to file
2. The patent application stage: the preparation and filing of the patent application
3. The prosecution stage: the process of interaction with the U.S. Patent and Trademark Office until the patent is granted

## Stage 1: From conception to patent application

The conception is the mental image of the complete invention in the mind of the inventor. The Court of Appeals of the Federal Circuit has used the definition generated in 1 Robertson on Patents 532 (1890) and requoted it in Hybritech, Inc. v. Monoclonal Antibodies, Inc. 231 USPQ 81 (1986): conception is the "formation in the mind of the inventor of a definite and permanent idea of the complete and operative invention as it is hereafter to be applied in practice."

The conception should be recorded and the record signed by the inventor. The record of the conception should then be read and signed by an appropriate witness. (Record-keeping procedures and the reasons for them are covered in detail in Chapter 12.)

In the next step, to complete the act of invention, the inventor (or someone under the inventor's direction) must reduce the invention to practice. "Actual" reduction to practice generally means physically carrying out the

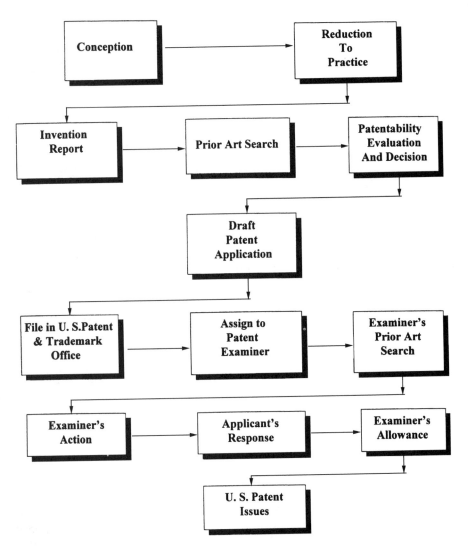

*Figure 6.1* Evolution of a patent — from conception to issuance.

conceived invention in the lab or workshop. The reduction to practice should be recorded and witnessed. It is also possible to fulfill the requirement, without physically reducing the invention to practice, by the act of filing a patent application that claims the invention. This approach is referred to as "constructive" reduction to practice.

Following the actual reduction to practice of the invention, most inventors will seek the help of a patent practitioner (i.e., a patent attorney or agent). Typically, a researcher in a corporate setting will do this by summarizing the details of the invention in an "Invention Report" and submitting it to the corporation's Patent Department.

The patent practitioner, whether in a corporate Patent Department or in an outside patent law firm, will generally order a prior art search to determine if the invention is novel and to review what has been published in the area of technology of the invention. Based on the information provided by the inventor and a study of the references found in the prior art search, the patent practitioner will make a patentability evaluation. The evaluation is a legal opinion, based to a great extent on scientific or technical information. If it is found that the invention is not patentable, the path toward a patent stops there. A trade secret approach might be considered at this point.

If the patentability evaluation is positive — that is, if in the opinion of the patent practitioner, the invention is patentable — the next step is a business decision. Typically, the decision will take three possibilities into account:

1. File a patent application
2. Maintain the invention as a trade secret
3. Defensively publish (i.e., make the invention public so that others will be barred from patenting it)

If a U.S. patent application is filed, the other two approaches (i.e., the trade secret approach or defensive publication) are still possible as long as the application is still pending and no publication has taken place. Both approaches will be considered in more detail later in this book. For now, to follow the stages in the evolution of a patent, it will be assumed that the patentability evaluation was positive and business considerations resulted in a decision to file a patent application.

## Stage 2:  The patent application

Basic to the philosophy of patents is that, in return for the patent, the inventor must fully disclose the invention. In so doing, the disclosure becomes a part of the technical literature and "promotes the progress of the useful arts." The full disclosure of the invention ensures that the public will have possession of the invention and how to make and use it after the patent expires.

The essential elements of a patent application are:

• A description of the invention
• One or more claims
• An oath of declaration signed by the inventor
• The payment of a filing fee

A patent is a hybrid document; it is both a legal and a technical document. Its preparation is a collaborative effort of a technical expert (the inventor) and a legal expert (the patent practitioner). Details of the contents of a patent will be discussed in depth in Chapter 7.

The descriptive portion of the patent application and the subsequently issued patent is called the "specification." It is a description required by statute (35 USC 112) to be written in sufficient detail to enable those skilled in the art to which it pertains to make and use the invention. For example, a specification of a patent (or patent application) on a chemical invention should be sufficiently clear and detailed to enable an ordinary chemist working in the same subject area to make and use the invention. Patents directed to mechanical and electrical inventions will commonly include drawings. Patents directed to chemical inventions will generally include working examples that can be written in "cookbook" style to guide other chemists in the practice of the invention. The specification concludes with one or more "claims" that particularly point out and distinctly claim the invention. The claims are statements that define the metes and bounds of the invention for legal purposes.

Once the inventor and the patent practitioner are satisfied that the specification and drawings, if any, properly *describe* the invention and how it can be practiced, and the claims properly *define* the invention, the inventor's declaration or oath can be signed. The addition of the required filing fee will then complete the application, and it can be submitted to the U.S. Patent and Trademark Office. A typical declaration form, to be signed by the inventor, is shown in Figures 6.2 and 6.3. As an inventor, one must read carefully what one is submitting and sign it.

U.S. patent law requires that the inventor must apply for the patent even though the rights to the invention may have been assigned to someone else, such as the inventor's employer.

## Stage 3: The prosecution

Upon receipt in the U.S. Patent and Trademark Office, the application will be stamped with the date of receipt and inspected to determine whether formal requirements have been met. If the application appears satisfactory, a serial number will be assigned and the "patent pending" period will begin. During this period, the application file will be identified only by its serial number. With the aid of the PTO computer system, the location of the file and its movement throughout the PTO, along with hundreds of thousands of other applications, will be followed at all times. The application is held in secret by the PTO until a patent issues; then, all the documents become public.

Once the application is filed, no substantive changes in the specification are permitted. Minor errors, such as obvious typographical or spelling errors may be corrected, but no new matter can be added to the specification. The claims, however, can be amended. The final wording and scope of the claims may be the result of discussions and arguments between the applicant and the patent examiner.

Following the preliminary inspection and assignment of the serial number, the application will be sent to an Examining Group that specializes in

Please type a plus sign (+) inside this box →☐

PTO/SB/01 (12-97)
Approved for use through 9/30/00. OMB 0651-0032
Patent and Trademark Office; U.S. DEPARTMENT OF COMMERCE
Under the Paperwork Reduction Act of 1995, no persons are required to respond to a collection of information unless it contains a valid OMB control number.

| DECLARATION FOR UTILITY OR DESIGN PATENT APPLICATION (37 CFR 1.63) | Attorney Docket Number | |
|---|---|---|
| | First Named Inventor | |
| | **COMPLETE IF KNOWN** | |
| | Application Number | / |
| ☐ Declaration Submitted with Initial Filing  **OR**  ☐ Declaration Submitted after Initial Filing (surcharge (37 CFR 1.16 (e)) required) | Filing Date | |
| | Group Art Unit | |
| | Examiner Name | |

**As a below named inventor, I hereby declare that:**

My residence, post office address, and citizenship are as stated below next to my name.

I believe I am the original, first and sole inventor (if only one name is listed below) or an original, first and joint inventor (if plural names are listed below) of the subject matter which is claimed and for which a patent is sought on the invention entitled:

the specification of which
☐ is attached hereto          *(Title of the Invention)*
OR
☐ was filed on (MM/DD/YYYY) [          ] as United States Application Number or PCT International

Application Number [          ] and was amended on (MM/DD/YYYY) [          ] (if applicable).

I hereby state that I have reviewed and understand the contents of the above identified specification, including the claims, as amended by any amendment specifically referred to above.

I acknowledge the duty to disclose information which is material to patentability as defined in 37 CFR 1.56.

I hereby claim foreign priority benefits under 35 U.S.C. 119(a)-(d) or 365(b) of any foreign application(s) for patent or inventor's certificate, or 365(a) of any PCT international application which designated at least one country other than the United States of America, listed below and have also identified below, by checking the box, any foreign application for patent or inventor's certificate, or of any PCT international application having a filing date before that of the application on which priority is claimed.

| Prior Foreign Application Number(s) | Country | Foreign Filing Date (MM/DD/YYYY) | Priority Not Claimed | Certified Copy Attached? YES | NO |
|---|---|---|---|---|---|
| | | | ☐ | ☐ | ☐ |
| | | | ☐ | ☐ | ☐ |
| | | | ☐ | ☐ | ☐ |
| | | | ☐ | ☐ | ☐ |

☐ Additional foreign application numbers are listed on a supplemental priority data sheet PTO/SB/02B attached hereto:

I hereby claim the benefit under 35 U.S.C. 119(e) of any United States provisional application(s) listed below.

| Application Number(s) | Filing Date (MM/DD/YYYY) | |
|---|---|---|
| | | ☐ Additional provisional application numbers are listed on a supplemental priority data sheet PTO/SB/02B attached hereto. |

[Page 1 of 2]

Burden Hour Statement: This form is estimated to take 0.4 hours to complete. Time will vary depending upon the needs of the individual case. Any comments on the amount of time you are required to complete this form should be sent to the Chief Information Officer, Patent and Trademark Office, Washington, DC 20231.  DO NOT SEND FEES OR COMPLETED FORMS TO THIS ADDRESS. SEND TO: Assistant Commissioner for Patents, Washington, DC 20231.

*Figure 6.2* PTO declaration for patent application.

PTO/SB/01 (12-97)
Approved for use through 9/30/00. OMB 0651-0032
Patent and Trademark Office; U.S. DEPARTMENT OF COMMERCE
Under the Paperwork Reduction Act of 1995, no persons are required to respond to a collection of information unless it contains a valid OMB control number.

Please type a plus sign (+) inside this box → ☐

# DECLARATION — Utility or Design Patent Application

I hereby claim the benefit under 35 U.S.C. 120 of any United States application(s), or 365(c) of any PCT international application designating the United States of America, listed below and, insofar as the subject matter of each of the claims of this application is not disclosed in the prior United States or PCT International application in the manner provided by the first paragraph of 35 U.S.C. 112, I acknowledge the duty to disclose information which is material to patentability as defined in 37 CFR 1.56 which became available between the filing date of the prior application and the national or PCT international filing date of this application.

| U.S. Parent Application or PCT Parent Number | Parent Filing Date (MM/DD/YYYY) | Parent Patent Number (if applicable) |
|---|---|---|
|  |  |  |
|  |  |  |

☐ Additional U.S. or PCT international application numbers are listed on a supplemental priority data sheet PTO/SB/02B attached hereto.

As a named inventor, I hereby appoint the following registered practitioner(s) to prosecute this application and to transact all business in the Patent and Trademark Office connected therewith: ☐ Customer Number [_____] → Place Customer Number Bar Code Label here
OR
☐ Registered practitioner(s) name/registration number listed below

| Name | Registration Number | Name | Registration Number |
|---|---|---|---|
|  |  |  |  |
|  |  |  |  |

☐ Additional registered practitioner(s) named on supplemental Registered Practitioner Information sheet PTO/SB/02C attached hereto.

Direct all correspondence to: ☐ Customer Number or Bar Code Label [_____]    OR ☐ Correspondence address below

| Name |  |  |  |
|---|---|---|---|
| Address |  |  |  |
| Address |  |  |  |
| City |  | State | ZIP |
| Country |  | Telephone | Fax |

I hereby declare that all statements made herein of my own knowledge are true and that all statements made on information and belief are believed to be true; and further that these statements were made with the knowledge that willful false statements and the like so made are punishable by fine or imprisonment, or both, under 18 U.S.C. 1001 and that such willful false statements may jeopardize the validity of the application or any patent issued thereon.

**Name of Sole or First Inventor:**      ☐ A petition has been filed for this unsigned inventor

| Given Name (first and middle [if any]) | Family Name or Surname |
|---|---|
|  |  |

| Inventor s Signature |  |  | Date |
|---|---|---|---|
| Residence: City | State | Country | Citizenship |
| Post Office Address |  |  |  |
| Post Office Address |  |  |  |
| City | State | ZIP | Country |

☐ Additional inventors are being named on the ____ supplemental Additional Inventor(s) sheet(s) PTO/SB/02A attached hereto

[Page 2 of 2]

*Figure 6.3*   Second page of PTO declaration form.

the technology to which the invention relates. Upon receipt by the Examining Group, it will be assigned to an individual examiner whose job it is to determine if the application meets the requirements to be awarded a patent. The examiner will take the application in its turn and will generally start by conducting his/her prior art search. The examiner will also have the benefit of the applicant's prior art search, since the applicant is required to inform the examiner of any known pertinent references, called an Information Disclosure Statement. The examiner will study the application in the light of the prior art and then issue an "office action" rejecting or allowing the claims of the application. In most instances, the first office action will be a rejection, which may be based on formalities or on the examiner's opinion that the claimed invention is shown in the prior art or is obvious from the prior art. The applicant may then respond by amending the claims to overcome the examiner's rejection and/or by presenting arguments refuting the examiner's rejection. This response must fall within a time period established by the PTO. The application will then be reconsidered in the light of the applicant's response and a second office action will be issued. Most commonly, the second action will be a "final" action from the examiner; that is, either an allowance or a final rejection. On receiving a final rejection, the applicant has the right to appeal to the Patent Office Board of Appeals and Interferences; the Board will review the case and render an opinion affirming or reversing the examiner's rejection.

The "patent pending" process, from filing to allowance, requires an average time of approximately 18 to 24 months, if the process goes smoothly. If there are problems, such as the need to appeal, the process may take considerably longer.

Following receipt of the allowance document from the patent examiner, the time has arrived for the payment of the Issue Fee. When this has been paid, the applicant will be notified as to when the patent will be issued, and its number. The formal copy of the patent will be sent to the applicant or his agent. Copies can be ordered, and the file of all written proceedings between the inventor, his agent, and the PTO will become public property and available to the public at the Patent Office in Crystal City, Arlington, VA.

## Trade secret approach

In the process outlined above, it was assumed that the invention was patentable and that business considerations dictated a patent approach to protect the invention. Because every invention starts its life as a trade secret and remains a trade secret until it is disclosed, the option of retaining the invention as a trade secret is an alternate approach to filing a patent application.

A trade secret is defined as information to be held in private, among a limited number of people, to be used to gain a competitive advantage in business. A problem, or disadvantage, is keeping the invention a secret. Another disadvantage to a trade secret is the ability of others to "invent around" the trade secret, or through "back engineering" discover the trade secret. The protection afforded a trade secret is granted *not* by federal law but perhaps by state law, if such laws exist. It should be noted that many of the state laws are directed at illegal methods of obtaining the inventor's secret, not the protection of the invention.

Many inventions have been and are currently protected as trade secrets. With trade secrets, there are no time limits for protection, as there are with the 20-year period in patents. In the patent approach, the invention must fit one of the statutory classes necessary for patenting, while as a trade secret, that is not necessary. Also, the governmental fees for a patent need not be spent, as no formal costs are incurred in the trade secret area. So, whether the invention is patentable or not, the trade secret approach can be followed.

## Defensive publication approach

The third approach to protection of the invention is *defensive publication.* This approach is simply to make the invention known to the public, typically through the publication of the invention in a scientific journal or through advertising. This approach "donates" the invention to the public but "prohibits" anyone else from getting a patent on the invention. The invention becomes prior art in the patent field. A great disadvantage is the fact that by using this approach, other people can now use the invention. While the cost of this approach is negligible, no protection has been obtained for the invention.

## Provisional application

With the 1995 GATT implementing legislation, a new category of U.S. patent application was created — a provisional application. A provisional application requires only the filing of a specification (and drawings, if necessary for an understanding of the invention) but does not require claims. A provisional application will be automatically abandoned after 1 year. It *cannot* mature into a patent; it will *not* be examined; but it *can* serve as a basis for priority for a complete application filed later. This application does not trigger the start of the 20-year patent term. To take advantage of the provisional application, a complete application must be filed within a year.

The provisional application has several important benefits. It has minimum legal and formal requirements. (The filing fee is only $150 for large businesses or $75 for small businesses and independent inventors.) Thus, it provides a mechanism whereby applicants can quickly and inexpensively establish an early effective filing date. The filing of a provisional application also provides up to 12 months to develop the invention, determine marketability, acquire funding or capital, seek licensing, and/or prepare for manufacturing.

If a provisional application is filed, and it is desired, during the 1-year period, to proceed with a patent application, the formal requirements for a "normal" patent application must be followed.

# An important factor: COST

A critical factor of all evaluations regarding the invention will be the cost. Whether the inventor is an independent inventor or a corporate employee, the cost affects the decision. It may simply be more than could be gained by its use. The cost may be too high for an individual inventor or a small company.

What fees will have to be paid to the U.S. Government? Table 6.1 shows the fees charged from 1991 to the present and how they have increased.

The fees are reviewed and revised on a 2-year cycle. Note carefully that an independent inventor or a small business is charged only 50% of the fees charged a large corporation. The initial fee to the government, payable with the application, will be $760 (or $380 if an independent inventor). The cost of the preparation of the application and prosecution is not included.

The current fees, as noted in Table 6.1, are normally suggested to Congress based upon the present fee and suggest changes due to the cost-of-living adjustments. This has been the generally accepted procedure. However, in 1998, Congress determined, after much lobbying by small inventor organizations and other groups, that the PTO was charging too much, and enacted legislation to reduce the fees. The fees had been based on legislative urging that the PTO become a government agency that supported its operations on the fees collected. After much discussion, the fees were reduced, in certain areas, from the preceding schedule.

Maintenance fees are also shown in Table 6.1. These are fees collected by the PTO after the issuance of the patent to keep the patent in force for the term of 20 years. The first such payment is due 3½ years after issuance, the next at 7½ years, and the last at 11½ years. If the patentee decides not to pay the maintenance fee when it is due, the patent expires.

The subject of total cost of the patent has not been discussed and will depend on the category into which the inventor falls. A small-entity company or a single inventor can get the fees due the Patent Office reduced by 50%

*Table 6.1*   U.S. Patent Office Fees

|  | 1991 | 1993 | 1994 | 1996 | 1998[a] |
|---|---|---|---|---|---|
| Filing fee ($) | 630 | 710 | 730 | 770 | 760 |
| Issue fee ($) | 1050 | 1170 | 1210 | 1290 | 1210 |
| Maintenance fees ($) |  |  |  |  |  |
| 3½ years | 830 | 930 | 960 | 1020 | 940 |
| 7½ years | 1670 | 1870 | 1930 | 2050 | 1900 |
| 11½ years | 2500 | 2820 | 2900 | 3080 | 2910 |
| Totals ($) | 6680 | 7500 | 7730 | 8210 | 7720 |

*Note:* Fees are reduced by 50% for independent inventors and small business entities.

[a] As of November 10, 1998, a reduction in fees due to legislative enactment.

by filing the proper forms. The inventor working for a corporation will have the costs borne by the corporation, and these costs will be considered in the process of deciding whether to file or not to file. An independent inventor or a small company inventor will have to consider the costs of a patent practitioner who is hired to draft and prosecute the application. A recent survey of patent costs estimates the cost of a non-technical patent application to be in the range of $2500 to $6000, while a complex application could double the cost. Additional costs to be incurred include responses to Office Actions and other correspondence with the Patent Office. These estimated costs do not include the fees that will be due the Patent Office. The costs can multiply if the case is appealed to the Board of Appeals, as additional time will be needed by the practitioner to protect the inventor's interest.

It is believed that the best protection available to the inventor is through the process of filing a patent application and gaining approval with the issuance of a patent. The trade secret approach is too risky and can be easily compromised, with the inventor losing all benefits.

The cost of a patent is not cheap, but it can be worth much money to the independent inventor or the corporation. Remember: an issued patent is a valuable asset.

## chapter seven

# The patent document as technical literature

Why is a patent document important to the scientist or engineer? If he/she is the inventor, the pride of ownership and accomplishment is quite high. But if someone else is the inventor, why look at it? Patent documents are not fiction; they are a truthful demonstration of scientific and technical achievements. These documents are a valuable font of technical knowledge; they show the current and past state of technology; they show what others have already achieved and provide experimental information that others can avoid repeating; and they assist in showing the prior art of the area of investigation.

Another valuable aspect of the patent literature that is often overlooked is the technology of expired patents that belongs to the public. Upon expiration of a patent, all rights belong to the public, and one is free to use and exploit them. Many well-known products become very profitable because they have patent protection. Once the patent expires, anyone can use the teachings of the patent and manufacture the product. Expired patents claiming chemical processes can be of great interest to research personnel; they only need to look in the patent literature, evaluate the expired patents, determine what has worked before, and thus determine what can be avoided in current work.

We have read or been taught that the university scientist must "publish or perish." In academe, emphasis is placed on getting experimental results into academic journals to promote the research reputation of the university and to help the status of the investigators. Many of these articles detail inventions that could have been patented. Patenting of inventions has not been widely promoted in academic research, but this attitude is changing. Institutions have learned that patents can bring money, as well as prestige, to their school. Many universities have added patent personnel to their staff to gain patent coverage on the inventions. The patents can then be licensed or sold, gaining financial rewards for the school.

Academic training has influenced us to use the journal literature during our college careers, and this has led to the same pattern in our industrial life. It is unfortunate that we were not trained to look at the patent literature, as the amount of information that is present may be greater than what is in the journals, and it may be more recent and disclose more technically. The discoveries of Ziegler and Natta in the field of olefin polymerization did not appear in the general chemical literature until about 1960, while their patents were filed in 1953 and published in several countries in 1955. People following the patent literature found the work, were able to base their research on the examples given in the patents, and started to expand on these famous inventions long before their appearance in the chemical journals.

Is a patent application difficult to read and understand? Many people compare reading a patent document with the fine print on an insurance policy. This is not true. A patent is not written as an article for a technical journal; rather, it is written in a format that has been established and is generally followed. The U.S. Patent Office requires a complete application to be filed before it can be examined. The PTO has a set of requirements for what an ideal document should contain. Although these requirements do not have to be followed to the letter, the rules suggest what must be included and how the application should appear. These rules can be found in Volume 37 of the *Code of Federal Regulations* (37 CFR). The rules specify an order of presentation of the contents of all applications, which are:

- Title of the invention
- Cross references to related applications (if any)
- Background of the invention
  - Field of the invention
  - Description of related art
- Summary of the invention
- Brief description of drawings (if any)
- Description of the preferred embodiment
- Claims
- Abstract of the disclosure

One reason for establishing such a format is that the application will become the text of the patent when it is allowed. Therefore, by having the text in a somewhat standard format, the ease of preparing the complete patent will be simplified and produce a common format for all patents. This allows the reader to be able to compare patents more readily.

Each of the items in this format is further defined and guidance provided in the rules to clarify what the PTO requires, but it is at the discretion of the drafter of the application whether or not to insert the titles into the application. In order to better explain how each of these items is used, one can use an actual patent as an example. Figures 7.1, 7.2, and 7.3 show a complete patent.

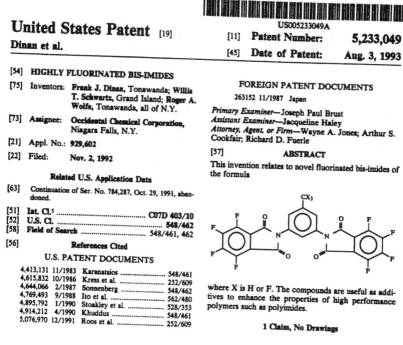

**United States Patent** [19]

**Dinan et al.**

|||||||| US005233049A

[11] **Patent Number:** **5,233,049**

[45] **Date of Patent:** **Aug. 3, 1993**

[54] **HIGHLY FLUORINATED BIS-IMIDES**

[75] Inventors: **Frank J. Dinan, Tonawanda; Willis T. Schwartz, Grand Island; Roger A. Wolfe, Tonawanda, all of N.Y.**

[73] Assignee: **Occidental Chemical Corporation, Niagara Falls, N.Y.**

[21] Appl. No.: **929,602**

[22] Filed: **Nov. 2, 1992**

**Related U.S. Application Data**

[63] Continuation of Ser. No. 784,287, Oct. 29, 1991, abandoned.

[51] Int. Cl.⁵ ............................................ C07D 403/10
[52] U.S. Cl. ........................................................ 548/462
[58] Field of Search ................................ 548/461, 462

[56] **References Cited**

**U.S. PATENT DOCUMENTS**

| 4,413,131 | 11/1983 | Karanatsios | 548/461 |
| 4,615,832 | 10/1986 | Kress et al. | 252/609 |
| 4,644,066 | 2/1987 | Sonnenberg | 548/462 |
| 4,769,493 | 9/1988 | Ito et al. | 562/480 |
| 4,895,792 | 1/1990 | Stoakley et al. | 528/353 |
| 4,914,212 | 4/1990 | Khuddus | 548/461 |
| 5,076,970 | 12/1991 | Roos et al. | 252/609 |

**FOREIGN PATENT DOCUMENTS**

263152 11/1987 Japan

*Primary Examiner*—Joseph Paul Brust
*Assistant Examiner*—Jacqueline Haley
*Attorney, Agent, or Firm*—Wayne A. Jones; Arthur S. Cookfair; Richard D. Fuerle

[57] **ABSTRACT**

This invention relates to novel fluorinated bis-imides of the formula

where X is H or F. The compounds are useful as additives to enhance the properties of high performance polymers such as polyimides.

**1 Claim, No Drawings**

*Figure 7.1* Title page of U.S. Patent 5,233,049.

Figure 7.1 is the first or **title page** of the patent. This page is not a part of the application but is a page assembled from data supplied by and to the PTO. Across the top of the page is the name of the country issuing the patent (United States Patent), the patent number (5,233,049), the last name of the inventors (Dinan et al.), and the date of issuance of the patent (August 3, 1993). This page will be described in further detail later in this chapter.

The majority of countries issuing patents have adopted a standard format for the first page of a patent. Depending on the country, the design and contents of the cover page may vary somewhat, but the similarity in general format will simplify the understanding of the patent.

Figure 7.2, the next page of the patent, states the **Title of the Patent or Title of the Invention**. Generally, the title should be short and as specific as possible. Related U.S. application data are recorded (if applicable).

Next comes the **Background of the Invention**. This section is divided into two parts: the field of the invention and the description of related art (prior art). The field of the invention is designed to allow the inventor to present what he/she believes is the scientific or technical area of the invention. The Description of the Related Art is the section where the applicant tells what is known about the area of the invention. A listing or brief discussion of the patents and other literature that relate to that area may be given. It is here that the applicant sets forth what he/she feels to be the

5,233,049

**1**

**HIGHLY FLUORINATED BIS-IMIDES**

This is a continuation of application Ser. No. 07/784,287, filed Oct. 29, 1991 now abandoned.

**BACKGROUND OF THE INVENTION**

The present invention relates to novel fluorine containing bis-imide compounds useful as additives to enhance the properties of high performance polymers such as polyimides.

Polyimide resins are used in a wide variety of industrial applications, based on their excellent thermo-oxidative stability, chemical stability, and dimensional stability of molded articles prepared from them. Fluorine-containing polyimides have been found particularly useful for applications requiring low moisture adsorption and high thermal stability.

It is known that the flammability of various material, especially polymeric materials, may be reduced by the incorporation therein of halogen-containing compounds. Various halogenated organic compounds have been useful as fire retardant additives for one type of resin but unsuitable for others because of incompatibility with a particular resin, or inability to withstand particular processing conditions, or at high temperatures. The fluorinated bis-imides of the present invention are particularly useful as additives to polyimide resins to serve as plasticizers and to enhance the already excellent fire retardant properties of such resins.

The incorporation of fluorine containing imide additives into polyimides to enhance the electrical properties, especially to lower the dielectric constant, is shown in U.S. Pat. No. 4,895,972. The patent teaches the incorporation of diamic acid additives, including fluorine containing diamic acids into polyamic acid resins prior to imidization.

U.S. Pat. No. 4,625,832 discloses the incorporation of fluorine containing phthalimides in combination with alkali metal salts, into thermoplastic, branched, aromatic polycarbonates to impart flameproofing properties thereto.

**SUMMARY OF THE INVENTION**

This invention provides novel fluorine-containing bis-imide compounds of the formula

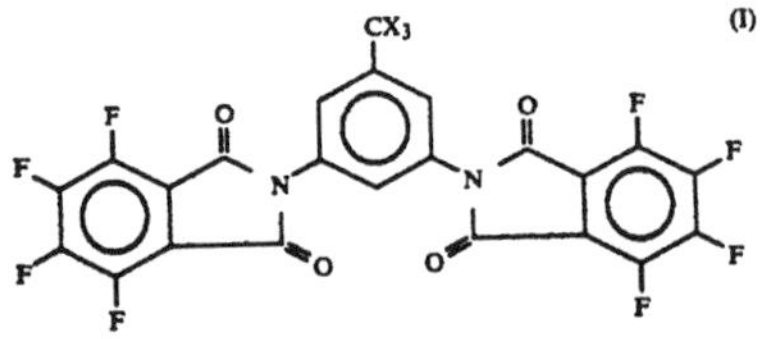

(I)

where X is H or F.

The fluorinated bis-imides of this invention are useful as flame retardant additives for various polymeric materials, especially polyimides and polycarbonates. For such purposes, the bis-imide additives are typically incorporated into the polymer in amounts of about 1.0 to about 30 percent by weight, based on the weight of the polymer. Polyimides are generally known to exhibit

**2**

desirable fire-resistant properties. However, for applications where it is desired to further enhance the already excellent fire-resistant properties, the fluorinated bis-imides of the present invention may be incorporated in the polyimide resins.

The fluorinated bis-imides of this invention may also be employed, optionally, in combination with an alkali metal salt, such as tripotassium or trisodium hexafluoroaluminate, to impart flame proofing characteristics to polycarbonates. The bis-imide compounds of this invention may also be used as plasticizers with various polymers, especially polyimides.

Bis-imides of the present invention as shown in Formula I above, wherein X is hydrogen, are prepared by the reaction of one mold of 3,5-diaminotoluene with two moles of tetrafluorophthalic anhydride. The bis-imides wherein X is fluorine, are prepared by the reaction of two moles of tetrafluorophthalic anhydride with one mole of 3,5-diaminobenzotrifluoride of the formula

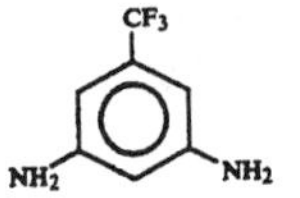

The reaction may be conveniently carried out by mixing the reactants in a solvent such as glacial acetic acid and heating to reflux for a period of time sufficient to complete the reaction, typically, three to four hours. Alternatively, the reaction may be carried out in a polar aprotic solvent, at a lower temperature, such as about 15°–30° C., to form the corresponding bis-amic acid which may then be employed as an additive to polyamic acid solutions prior to imidization.

The following examples are provided to further illustrate the present invention and the procedure by which the bis-imides of this invention may be prepared.

**EXAMPLE I**

A mixture of diaminobenzotrifluoride (1.0 g; 0.0057 mole) and tetrafluorophthalic anhydride (3.6 g; 0.0167 mole) in 30 mL of glacial acetic acid was heated to reflux and maintained thereat for about 3 hours. The reaction mixture was then cooled to room temperature, filtered, and teh solids washed with acetic acid, then hexane and vacuum dried at 100° C. to yield 2.8 g (85%) of the bis-imide of diaminobenzotrifluoride and tetrafluorophthalic anhydride (m.p. 331°–332° C.). The structure and purity (99.4%) was confirmed by GC/MS.

**EXAMPLE II**

A mixture of diaminotoluene (0.75 g; 0.0061 mole) and tetrafluorophthalic anhydride (4.10 g; 0.0186 mole) in 30 mL of glacial acetic acid was heated to reflux and

*Figure 7.2*    Second page of U.S. Patent 5,233,049.

dominant prior art in the area. This section should explain the problem involved and indicate that the problem is solved by this invention.

The next section is the **Summary of the Invention**, which is a brief summary or a general statement of the invention. Here, the applicant can

## 5,233,049

**3**

maintained thereat for about 3.5 hours. The reaction mixture was then cooled to room temperature, filtered, and the solids washed with acetic acid, then hexane and vacuum dried at 100° C. to yield 2.52 g (77.9%) of the bis-imide of diaminobenzotrifluoride and tetrafluorophthalic anhydride (m.p. 310°-311° C.). The structure and purity (98.8%) was confirmed by GC/MS.

What is claimed is:

1. A bis-imide of the formula

**4**

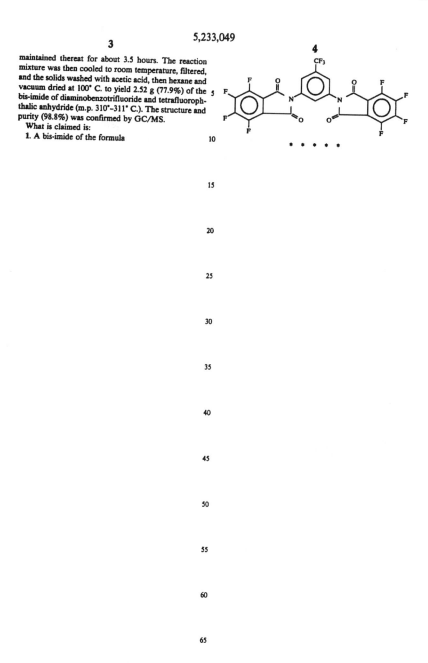

5

10

• • • • •

15

20

25

30

35

40

45

50

55

60

65

*Figure 7.3*    Third page of U.S. Patent 5,233,049.

state the advantages and objectives of the invention and how it solves the problems previously present in the art. In chemical applications, it may set out, in general terms, the utility of the invention.

The next section is a **Brief Description of the Drawings** (if any), and states very concisely what the drawing is or shows. Drawings are used with the written description of the invention to clarify the details of the invention and simplify the text.

The next section, **Description of the Preferred Embodiments,** or **Detailed Description of the Invention,** is the main text of the application. The rules require the specification to include a written description of the invention or discovery and of the manner and process of making and using the same. It is required to be in such full, clear, concise, and exact terms as to enable any person, skilled in the art, to make and use the invention.

If there are drawings, the Detailed Description will generally include a detailed explanation of them. Patents on chemical inventions generally do not include drawings but do include examples, often written in cookbook form, to show how to practice the invention.

Additionally, the *best* procedure for carrying out the invention known at the time of filing of the patent application must be given — and not just any procedure. If the best procedure is not presented, the patent is subject to revocation for failure to fulfill the requirement for the best procedure.

It is not necessary to explain why the invention functions or the theories behind the invention. Each drafter of an application has different concepts of how an application should be developed. Prior to presenting the Examples, it is generally advised to discuss the components that go into the invention and provide for alternative materials, compounds, and methods. This presents the position to the patent examiner that the invention is not limited to just one set of components, but is broader.

The function of a patent application is twofold. One function is to describe the invention so that it can be understood and practiced by someone skilled in the art. Second, it is to inform the public of the limits of the invention that are protected by the patent. The part of the application that informs the public of the limits of the patent is the Claims.

The **Claims** section of the application is one of the most difficult and important sections to write. The Claims determine the legal limits of the invention. Generally, they will claim a broader area than what is clearly defined by the patent. Claims will be further discussed in Chapter 10.

After the Claims comes the last section of the application, the **Abstract**. This does not follow the Claims in Figure 7.3, but is moved by the PTO to be placed on the Title Page of the Patent. This Abstract is different from the Summary of the Invention, as this is an abstract of the disclosure. This means that the contents of the abstract should be such as to enable the reader to ascertain the character of the subject matter covered by the technical disclosure and should include what is new in the art to which the invention pertains. It does not define the legal scope of the patent.

Returning to the Title Page (Figure 7.1), it was earlier noted that the issuing country, patent number, and inventor's name are listed across the top of the page. Notice that behind some of these items, or in front of them, are small boxed numbers (e.g., [19]). These are the code numbers that have been established for identifying bibliographic information. This code, INID, an acronym for Internationally agreed Numbers for the Identification of Data, has been created to define important dates and data and appears on the title sheet of every patent. The INID code has been adopted by all countries in the world as a method of having a standardized system to identify data. Later in this chapter, these code numbers will be used to explain their use in foreign patents. On the first or title page of this patent, these numbers indicate the following data:

[54]  This is the official title of the patent document and is, by the rules of the U.S. Patent Office, to be brief and descriptive of the invention.

[75]  This is the name of the inventor or inventors. In U.S. patents, the name and residence are printed on the document.

[73]  This is the name and address of the assignee or owner of the patent, at the time of issuance.

[21]  This is the serial number of the application.

[22]  This is the filing date of the application.

[63]  This indicates an earlier filing, showing number and date, from which this application is a continuation.

[51]  This is the international classification of the patent.

[52]  This is the U.S. classification of the patent.

[58]  This is a listing of the class and subclasses searched by the patent examiner.

Items 51, 52, and 58 are all concerned with classification of the patent. These are methods that index the patent into a system of classes and subclasses that divide the technology into smaller and more specific indexing areas. This will be discussed in more detail in Chapter 8.

[56]  These are the references cited by the examiner during the examination of the application. This area is divided into several parts: the U.S. patents cited, foreign patents cited, and other publications. Also listed in this section are the names of the examiner who conducted the examination of the application, and the attorney, agent, or firm who represented the applicant.

[57]  This is an abstract of the patent.

The final item is a listing of the number of claims and drawings. (In the event drawings or circuit diagrams were included in the application, a drawing will be reproduced on the lower portion of the cover page.)

Patents vary in length, and this is one of the shorter ones. The amount of data, experiments, theories, and description contained in a patent varies with the drafter of the application and the amount of data the inventor possesses. Length or brevity of a patent has no effect on its quality or contents. Do not dismiss a patent as being of little importance just because it is short. Study and evaluate it.

### –International agreed Numbers for the Identification of Data [INID]

[10] **Document Identification.**
[11]  Number of the document
[12]  Plain language, designation of the kind of document.
[13]  Kind of document code according to WIPO Standard ST. 16.
[19]  WIPO Standard ST. 3 code or other identification, of the office publishing the document.

[20] **Domestic Filing Data**
[21]  Number(s) assigned to the application(s).
[22]  Date(s) of filing application(s).
[23]  Other date(s), including date of filing complete specification following provisional specification and exhibition filing date.
[24]  Date from which industrial property rights may have effect.
[25]  Language in which the published application was originally filed.
[26]  Language in which the application is published.

[30] **Priority Data**
[31]  Number(s) assigned to priority application(s).
[32]  Date(s) of filing of priority application(s).
[33]  WIPO Standard ST.3 Code identifying the national patent office allotting the priority application number or the organization allotting the regional priority application number for the international applications filed under the PCT, the Code "WO" is to be used.
[34]  For priority filings under regional or international arrangements, the WIPO Standard ST.3 Code identifying at least one country party to the Paris Union for which the regional or international application was made.

[40]  **Date(s) of making available to the public**
[41]  Date of making available to the public by viewing, or copying on request, an unexamined document, on which no grant has taken place on or before the said date.
[42]  Date of making available to the public by viewing, or copying on request, an examined document, on which no grant has taken place on or before the said date.
[43]  Date of publication by printing or similar of an unexamined document, on which no grant has taken place on or before the said date.
[44]  Date of publication by printing or similar process of an examined document on which no grant has taken place on or before the said date.
[45]  Date of publication by printing or similar process of an examined document on which grant has taken place on or before the said date.
[46]  Date of publication by printing or similar process of the claim(s) only of a document.
[47]  Date of making available to the public by viewing, or copying on request, a document on which grant has taken place on or before the said date.

[50]  **Technical Information**
[51]  International Patent Classification.
[52]  Domestic or national classification.
[53]  Universal Decimal Classification.
[54]  Title of the Invention.
[55]  Keywords.
[56]  List of prior art documents. If separate from descriptive text.
[57]  Abstract or claim.
[58]  Field of search

*Figure 7.4*   Page 1 of INID coding system numbers for patents.

The INID coding system, described earlier, is very useful in understanding foreign patents, as their language may not be easily read by most people. A brief listing of this code is shown in Figures 7.4 and 7.5. Using this code, a person can look at a patent, in a language unknown to the reader, and determine who the inventor is, the filing date, and other pertinent information. Figure 7.6 shows the title page of a European Patent Office (EPO) Patent;

**[60]** **References to other legally related domestic patent documents including unpublished applications therefore**

[61] Number and, if possible, filing date of the earlier application, or number of the earlier publication, or number of earlier granted patent, inventors' certificate, utility model or the like to which the present document is an addition.

[62] Number and, if possible, filing date of the earlier application from which the present document has been divided out.

[63] Number and filing date of the earlier application of which the present document is a continuation.

[64] Number of the earlier publication which is "reissued."

[65] Number of a previously published patent document concerning the same application.

[66] Number and filing date of the earlier application of which the present document is a substitute, i.e., a later application filed after the abandonment of an earlier application for the same invention.

**[70]** **Identification of parties concerned with the document**

[71] Name(s) of application(s)

[72] Name(s) of inventor(s) if known to be such

[73] Name(s) of grantee(s)

[74] Name(s) of attorney(s) or agent(s)

[75] Name(s) of inventor(s) who is (are) also applicant(s)

[76] Name(s) of inventor(s) who is (are) also applicant(s) and grantee(s)

**[80]** **Identification of data related to International Conventions other than the Paris Convention**

[81] Designated state(s) according to the PCT

[82] Elected state(s) according to PCT.

[83] Information concerning the deposit of microorganisms, e.g. under the Budapest Treaty.

[84] Designated contracting states under regional patent conventions.

[85] Date of fulfillment of the requirements of Articles 22 and/or 39 of the PCT for introducing the national procedure according to the PCT.

[86] Filing date of the regional or PCT application, i.e. application filing date, application number, and, optionally, the language in which the published application was originally filed.

[87] Publication data of the regional or PCT application, i.e. publication date, publication number, and, optionally the language in which the application is published.

[88] Date of deferred publication of the search report.

[89] Document number and country of origin of the original document according to the CMEA Agreement on Mutual Recognition of Inventors' Certificates and other Titles of Protection for Inventions.

*Figure 7.5* Page 2 of INID coding system numbers for patents.

Europäisches Patentamt

(19)    European Patent Office

Office européen des brevets

(11)    **EP 0 812 233 B1**

(12)    **EUROPÄISCHE PATENTSCHRIFT**

(45) Veröffentlichungstag und Bekanntmachung des Hinweises auf die Patenterteilung:
**30.09.1998 Patentblatt 1998/40**

(51) Int. Cl.[6]: **B01D 5/00**, B01D 3/42, B01D 3/06, B01D 1/30, D06L 1/10, D06F 43/08

(21) Anmeldenummer: 96907333.7

(22) Anmeldetag: 01.03.1996

(86) Internationale Anmeldenummer:
PCT/EP96/00859

(87) Internationale Veröffentlichungsnummer:
WO 96/26780 (06.09.1996 Gazette 1996/40)

(54) **VERFAHREN UND VORRICHTUNG ZUR WIEDERAUFBEREITUNG EINES VERUNREINIGTEN LÖSEMITTELS**

PROCESS AND DEVICE FOR REGENERATING A CONTAMINATED SOLVENT

PROCEDE ET DISPOSITIF DE REGENERATION D'UN SOLVANT CONTAMINE

(84) Benannte Vertragsstaaten:
AT BE CH DE FR GB IT LI LU NL SE

(30) Priorität: 01.03.1995 DE 19507126
18.05.1995 DE 19518346
13.10.1995 DE 19538214

(43) Veröffentlichungstag der Anmeldung:
17.12.1997 Patentblatt 1997/51

(73) Patentinhaber:
Baumann, Didda Maria Janina
97076 Würzburg/Lengfeld (DE)

(72) Erfinder: BAUMANN, Walter
D-99819 Elsenach (DE)

(74) Vertreter:
Bell, Hans Christoph, Dr. et al
Hansmann & Vogeser,
Patent- und Rechtsanwälte,
Postfach 80 01 40
65901 Frankfurt (DE)

(56) Entgegenhaltungen:
EP-A- 0 021 021        EP-A- 0 236 813
EP-A- 0 284 341        WO-A-93/24198
US-A- 2 881 116

EP 0 812 233 B1

Anmerkung:    Innerhalb von neun Monaten nach der Bekanntmachung des Hinweises auf die Erteilung des europäischen Patents kann jedermann beim Europäischen Patentamt gegen das erteilte europäische Patent Einspruch einlegen. Der Einspruch ist schriftlich einzureichen und zu begründen. Er gilt erst als eingelegt, wenn die Einspruchsgebühr entrichtet worden ist. (Art. 99(1) Europäisches Patentübereinkommen).

*Figure 7.6*    Title page of European Patent EP 0 812 233 B1.

CESKA A SLOVENSKA
FEDERATIVNI
REPUBLIKA
(19)

FEDERALNI URAD
PRO VYNALEZY

# PATENTOVÝ SPIS

(21) Číslo příhlášky : 2705-90

(22) Přihlášeno : 01.06.90

(30) Prioritni data :

(40) Zveřejněno : 17.12.91

(47) Uděleno : 30.09.92

(24) Oznámeno udělení ve Vöstniku: 18.11.92

(11) Číslo dokumentu :
## 277101

(13) Druh dokumentu : BE
(51) Int. Cl. [5] :
    B 21 B 27/06

(73) Majitel patentu : Vítkovice, s.p., Ostrava, CS

(72) Puvodco vynálezu : Tučník Drahomír, Ostrava, CS;
Myška Josef ing., Ostrava, CS

(54) Název vynálezu : Chladicí zařízení pracovních válců tažné stolice
při plynulém odlévání kovu

(57) Anotace :

Zařízení sestává z rotacních rozvádecích objímek /1, 2/, nasazenych na hridele /3, 4/ pracovních válcu /5, 6/, do nichz je napojeno jak přívodní potribí /19/, tak i odpadní potrubí /20/ chladicího media. Vstupní vnitřní prostory (21) a vystupní vnitřní prostory /24/ rotacních rozvádécích objímek /1, 2/ jsou propojeny kanálky /22, 25/ se stredovym slepym otvorem a s mezikruhovym vnitrním prostorem, které jsou vytvořeny v hřídelích /3, 4/. Přívodní potrubí /19/ a odpadní potrubí /20/ z rotacní rozvádecí objímky /1/ hřídele /3/ horního pracovního válce (5) je propojeno s rotačni rozvádecí objímkou /12/, nasazenou na čepu /11/, která je napojena na centrální rozvod chladicího média.

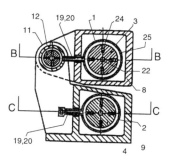

*Figure 7.7* Title page of Slovenian Patent 277101.

Figure 7.7 shows the title page of a Slovak Republic patent; and Figure 7.8 shows the title page of a Romanian patent. From these documents, one can quickly see the use and value of this coding system.

Many people feel that patents are difficult to read and that it is difficult to locate patents that apply to their area of research. Neither is correct. As previously stated, a patent is written in a specific form and uses language that is not difficult to understand. It *must* explain the invention so that any person skilled in this area of science can understand and reproduce the results. This is one way that a patent differs from a journal article. A journal article is designed to explain one factor of an experiment and does not have to disclose the best method or state its utility.

How does one find the patents that are pertinent to a particular field of interest, and where can they be found? First, one must define the field of

| ROMANIA | BREVET DE INVENŢIE (19) RO(11) 104139 |
|---|---|
| OFICIUL DE STAT PENTRU INVENTII SI MARCI | (12)DESCRIEREAINVENŢIEI |

(21)　Cerere dé brevet nr: **139829**
(22)　Data inregistrării: **13.05.89**
(61)　Complementară la inventia
　　　brevet　nr:
(45)　Data publicării: **15.12.93**

(51)Int. Cl.⁴: **B 27 D 1/10**

(86)　Cerere internatională (PCT)
　　　nr:　data:
(87)　Publicarea cererii internationale
　　　nr:　data:
(89)

(30)　Prioritate:
　　　(32)　Data:
　　　(33)　Tara:
　　　(31)　Certificat nr:

(71)　Solicitant; (73) Titular: Întreprinderea de Scule, Râsnov, judeţul Braşov
(72)　Inventator: ing. Rogalschi Eugen, Râşnov, judeţul Braşov

(54)　Dispozitiv electronic cu program de comandă a operatiilor
de îmbinare şi debitare a furnirului

(57)　**Rezumat**

Invenţia are ca obiect un dispozitiv programator electronic de comanda a operatillor de îmbinare şi debiltare a furnirului, format dintr-un traductor incremental de rotaţie, un comutator electronic cu două programe, ce comandă două fotorelee, şi un bloc de comandă şi coordonare a comenzilor.

(19)RO(11)104139

Preţ lei 832.00

Grupa 7

*Figure 7.8*　Title page of Romanian Patent 104,139.

interest and the area of that field one wants to study. One knows from using any of the technical abstract journals that if one more narrowly and carefully defines the area being sought, then the more pertinent references will be found. It is the same in the patent field. Patents are indexed in classes and subclasses, and an index volume exists that can help find the location for the search. The narrowing of the area will lead to finding the most information in the shortest time. The classification system will be discussed in detail in Chapter 8.

Where does one find the patents? Naturally, in the U.S. Patent and Trademark Office in Arlington, VA. There in the Public Search Room, all of the patents are available and grouped into classes and subclasses. Basically, the patents are placed in a class that defines a technical area. Each class is then subdivided into subclasses that more carefully define the inventive area. This allows the searcher to find them quickly and easily. All of the patents classified into these subclasses are stored on shelves in the Search Room. This allows the searcher to read them at a leisurely pace. Copies can be ordered or made, as one wishes. If the Washington, D.C. area is not convenient for a search, the Patent Office has established a Patent Depository Library (PDL) system that maintains branch patent files in libraries throughout the U.S. These libraries are continually being established, and a recent list of PDLs is given in Figure 7.9. Many of these are located at colleges and universities; others at public and state libraries. Each has copies of patents, and their files are constantly updated. There are also commercial databases that can be searched by computer.

With millions of patents, most of them cross-referenced into different areas, the need for locating and searching this mass of documents is an ideal area for automation, and the PTO has been working on these problems for many years. The problem was realized in the 1920s and attempts at solution were made; but it was not until the advent of computer technology that a feasible approach to the problem was found. The PTO is actively working on the problem and has made progress. It has established its own databases and programs to search the patents and their contents. The system, utilizing a keyword method, is being used by the examiners and patent searchers.

Chapter 9 will discuss, in depth, the use of computers to search the patent files and its literature. Like all systems of data retrieval, there are problems; these are discussed in Chapters 8 and 9.

A common problem facing the computer age is terminology and spelling. As an example, a computer search for a "color coupler dye" for photography will result in many responses. It will not uncover "*colour* coupler dyes" because the computer does not "think." It cannot realize that color and colour are just different spellings of the same word. Spelling depends on one's upbringing, education, or geographical location. The printed patents, regardless of spelling, are classified in the same location because they have been placed there by humans who know the different spellings. Just think of other words, like colour, with different spellings; this is a large problem. Also, if a word is misspelled in a patent, the computer will not correct the misspelling, but will just overlook the entry. This also creates problems.

The Patent and Trademark Depository Library Program consists of 83 Patent and Trademark Depository Libraries (PTDLs) in the 50 states, the District of Columbia, and Puerto Rico. Call ahead of your visit for hours of operation, services, and fees.

|  |  |  |
|---|---|---|
| Alabama | * Auburn University: Ralph Brown Draughon Library | 334-844-1747 |
| | Birmingham Public Library | 205-226-3620 |
| Alaska | Anchorage: Z. J. Loussac Public Library, | 907-562-7323 |
| | Anchorage Municipal Libraries | |
| Arizona | * Tempe: Noble Science and Engineering Library, | 602-965-7010 |
| | Arizona State University | |
| Arkansas | * Little Rock: Arkansas State Library | 501-682-2053 |
| California | * Los Angeles Public Library | 213-228-7220 |
| | Sacramento: California State Library | 916-654-0069 |
| | San Diego Public Library | 619-236-5813 |
| | * San Francisco Public Library | 415-557-4500 |
| | ** Sunnyvale Center for Innovation, Invention & Ideas | 408-730-7290 |
| Colorado | Denver Public Library | 303-640-6220 |
| Connecticut | Hartford Public Library | 860-543-8628 |
| | New Haven Free Public Library | 203-946-8130 |
| Delaware | Newark: University of Delaware Library | 302-831-2965 |
| District of Columbia | Washington: Founders Library, Howard University | 202-806-7252 |
| Florida | * Fort Lauderdale: Broward County Main Library | 954-357-7444 |
| | * Miami-Dade Public Library | 305-375-2665 |
| | Orlando: University of Central Florida Libraries | 407-823-2562 |
| | Tampa Campus Library, University of South Florida | 813-974-2726 |
| Georgia | Atlanta: Library & Information Center, Georgia Institute of Technology | 404-894-4508 |
| Hawaii | * Honolulu: Hawaii State Library | 808-586-3477 |
| Idaho | Moscow: University of Idaho Library | 208-885-6235 |
| Illinois | Chicago Public Library | 312-747-4450 |
| | Springfield: Illinois State Library | 217-782-5659 |
| Indiana | Indianapolis-Marion County Public Library | 317-269-1741 |
| | West Lafayette: Siegesmund Engineering Library, Purdue University | 765-494-2872 |
| Iowa | Des Moines: State Library of Iowa | 515-281-4118 |
| Kansas | * Wichita: Ablah Library, Wichita State University | 316-978-3155 |
| Kentucky | * Louisville Free Public Library | 502-574-1611 |
| Louisiana | Baton Rouge: Troy H. Middleton Library, Louisiana State University | 504-388-8875 |
| Maine | Orono: Raymond H. Fogler Library, University of Maine | 207-581-1678 |
| Maryland | College Park: Engineering and Physical Sciences Library, University of Maryland | 301-405-9157 |
| Massachusetts | Amherst: Physical Sciences and Engineering Library, University of Massachusetts | 413-545-1370 |
| | * Boston Public Library | 617-536-5400 Ext. 265 |
| Michigan | Ann Arbor: Media Union Library, The University of Michigan | 313-647-5735 |
| | Big Rapids: Abigail S. Timme Library, Ferris State University | 616-592-3602 |
| | ** Detroit: Great Lakes Patent and Trademark Center, Detroit Public Library | 313-833-3379 |
| Minnesota | * Minneapolis Public Library & Information Center | 612-630-6120 |
| Mississippi | Jackson: Mississippi Library Commission | 601-359-1036 |
| Missouri | * Kansas City: Linda Hall Library | 816-363-4600 |
| | * St. Louis Public Library | 314-241-2288 Ext. 390 |
| Montana | Butte: Montana Tech of the University of Montana Library | 406-496-4281 |
| Nebraska | * Lincoln: Engineering Library, University of Nebraska-Lincoln | 402-472-3411 |
| Nevada | Reno: University Library, University of Nevada-Reno | 702-784-6500 Ext. 257 |
| New Hampshire | Concord: New Hampshire State Library | 603-271-2239 |
| New Jersey | Newark Public Library of Piscataway: | 973-733-7779 |
| | Library of Science and Medicine, Rutgers University | 732-445-2895 |

*Figure 7.9*   Listing of patent depository libraries in the United States.

| | | |
|---|---|---|
| New Mexico | Albuquerque: Centennial Science and Engineering Library, The University of New Mexico | 505-277-4412 |
| New York | Albany: New York State Library | 518-474-5355 |
| | * Buffalo and Erie County Public Library | 716-858-7101 |
| | New York: Science, Industry and Business Library, New York Public Library | 212-592-7000 |
| | Stony Brook: Engineering Library, State University of New York | 516-632-7148 |
| North Carolina | * Raleigh: D. H. Hill Library, North Carolina State University | 919-515-2935 |
| North Dakota | Grand Forks: Chester Fritz Library, University of North Dakota | 701-777-4888 |
| Ohio | Akron-Summit County Public Library | 330-643-9075 |
| | Cincinnati: The Public Library of Cincinnati and Hamilton County | 513-369-6971 |
| | * Cleveland Public Library | 216-623-2870 |
| | Columbus: Ohio State University Libraries | 614-292-6175 |
| | * Toledo/Lucas County Public Library | 419-259-5212 |
| Oklahoma | * Stillwater: Oklahoma State University | 405-744-7086 |
| Oregon | Portland: Lewis & Clark College | 503-768-6786 |
| Pennsylvania | * Philadelphia: The Free Library of | 215-686-5331 |
| | Pittsburgh: The Carnegie Library of | 412-622-3138 |
| | University Park: Pattee Library, Pennsylvania State University | 814-865-4861 |
| Puerto Rico | Mayaguez: General Library, University of Puerto Rico | 787-832-4040 Ext. 3459 |
| Rhode Island | Providence Public Library | 401-455-8027 |
| South Carolina | Clemson: R. M. Cooper Library, Clemson University | 864-656-3024 |
| South Dakota | Rapid City: Devereaux Library, South Dakota School of Mines and Technology | 605-394-1275 |
| Tennessee | Memphis & Shelby County Public Library & Information Center | 901-725-8877 |
| | Nashville: Stevenson Science and Engineering Library, Vanderbilt University | 615-322-2717 |
| Texas | Austin: McKinney Engineering Library, The University of Texas at Austin | 512-495-4500 |
| | * College Station: Sterling C. Evans Library, Texas A&M University | 409-845-3826 |
| | * Dallas Public Library | 214-670-1468 |
| | ** Houston: The Fondren Library, Rice University | 713-527-8101 Ext. 2587 |
| | Lubbock: Texas Tech University Library | 806-742-2282 |
| Utah | * Salt Lake City: Marriott Library, University of Utah | 801-581-8394 |
| Vermont | Burlington: Bailey/Howe Library, University of Vermont | 802-656-2542 |
| Virginia | * Richmond: James Branch Cabell Library, VA Commonwealth University | 804-828-1104 |
| Washington | * Seattle: Engineering Library, University of Washington | 206-543-0740 |
| West Virginia | * Morgantown: Evandale Library, West Virginia University | 304-293-2510 Ext. 5113 |
| Wisconsin | Madison: Kurt F. Wendt Library, University of Wisconsin-Madison | 608-262-6845 |
| | Milwaukee Public Library | 414-286-3051 |
| Wyoming | Casper: Natrona County Public Library | 307-237-4935 |

* Denotes APS-Text Access                          ** Denotes Partnership PTDL

PTDL program information is also found on the Internet at http://www.uspto.gov

*Figure 7.9 (continued)*

## chapter eight

# The basic principles
# of patent searching

U.S. patents are issued every Tuesday of the year and have sequential numbers. Currently, they are over 6 million patents, so it is impractical to search in a random manner. With this number of patents, some kind of system must be utilized. In 1830, legislation was passed that required the PTO to keep a record of the patents issued. Prior to this time, patents were issued, but no records were kept of the titles or who received the patents. After 1830, all patents were numbered. At that time, it was realized that a system was needed to separate the patents into categories so that patents belonging to a similar scientific area could be located. A system of classes and subclasses was started and is still in use today.

This system of classifying patents began with about a dozen classes, and each had a subclass or two. Now, there are about 400 classes with subclasses numbering about 160,000. Each class defines a portion of technology, with the subclasses dividing that technology into smaller and smaller, more specific divisions. A patent can be cross-referenced into several subclasses and even into other classes. The classes and subclasses are in a constant state of revision and updating. The classes change as technology and information change. Classes disappear (e.g., "Buggies"); classes appear (e.g., "Biotechnology"). Each patent, when it is issued, is placed into at least one class and subclass, and generally cross-referenced to others.

The notation one sees on the patents in this book and other literature places the class number first, followed by a slash and the number of the subclass. For example, 570/206 means class 570, subclass 206. This is merely a shorthand way of writing this information.

These classes and subclasses can be found in the *Manual of Patent Classification*, where each class is found with all of its subclasses. This is a looseleaf volume that is revised and updated three or four times a year. Another volume that is important to the searcher is the *Index to the Patent Classifications*. The *Index*, published annually, is an alphabetical list of subjects with about 60,000 entries referencing specific classes and subclasses in the classification system. This volume allows one to locate relevant classes and subclasses in the *Manual*. Another set of volumes that can be of assistance is the *Class and Subclass Definitions*. These volumes define each class and subclass and provide references to other classes and subclasses that are relevant to the subject area. Sample pages of these volumes follow. Figures 8.1, 8.2, and 8.3 show a complete class 570, while Figure 8.4 shows a representative page of the *Index*. Figure 8.5 shows a typical page of the *Class Definition* volume.

Searching is slow and tedious, and anything that can speed up the process is very helpful. Before starting a search, stop and consider how best to identify the area of search. Consider carefully, *what is the invention?* Consider several ways to describe the invention and write them down. Check each of them in the *Index* volume and see if they all lead to the same class and subclass. If they do not, perhaps it will be necessary to look in all of the referenced classes to complete the search. Remember, the classification system of the PTO is not an exact science; it is influenced by humans. One may have differences of opinion as to classification, but the cross-reference system will generally eliminate the errors.

There are different kinds of searches and each requires the searcher to work slightly differently, depending on the object of the search. There are five common searches, each having a different purpose and each a different approach.

1. **Patentability search**: This is a type of search in which one is seeking to learn what has been patented in the area of the concept. *This search will not determine the prior art of the area. That is not its purpose.* This type of search may be called by other names, including pre-examination, preliminary, patent novelty, etc. Generally, this type of search is limited in scope to U.S. patents. This is a negative type of search, as one is hoping **not** to find a patent identical to the concept being searched. This is a general search of the concept, idea, or discovery to learn what has been patented and its relation to the concept. This search is usually conducted prior to the determination of whether or not to file a patent application. Often, the scope of a patentability search is expanded by searching a broader area of the idea or concept to learn the state of the prior art in the general area, rather than limiting the search to a specific concept.

2. **Infringement search**: This is a type of search in which one wishes to determine if the claims of a patent infringe any other patent. What is desired is to determine if the claims of unexpired patents in the field of the invention overlap the claims of the patent in question. The search is limited to the active or unexpired patents. Infringement searches are also carried out to determine whether some activity, such as a new process or new product to be manufactured or sold, will infringe the claims of someone else's patent.

CLASS 570  ORGANIC COMPOUNDS -- PART OF THE CLASS 532-570 SERIES          570-1

DECEMBER 1996

> IN THIS SERIES OF CLASSES, CLASS 570 IS TO BE CONSIDERED AS AN INTEGRAL PART
> OF CLASS 260 AND FOLLOWS THE SCHEDULE HIERARCHY, RETAINING ALL PERTINENT
> DEFINITIONS AND CLASS LINES OF CLASS 260.

| | ORGANIC COMPOUNDS (Class 532, Subclass 1) | | |
|---|---|---|---|
| 101 | .HALOGEN CONTAINING | 135 | .....Unsaturated |
| 102 | ..With preservative or stabilizer | 136 | ......Fluorine is sole halogen |
| 103 | ...To prevent or reduce polymerization | 137 | .....Bromine or iodine containing |
| 104 | ....Nitrogen bonded directly to oxygen in preservative or stabilizer | 138 | ...Polymerization of unsaturated compound |
| 105 | ....Oxygen single bonded directly to benzene ring in preservative or stabilizer | 139 | ....With chain terminating agent (e.g., telogen, etc.) |
| 106 | ....Sulfur containing preservative or stabilizer | 140 | ...From organic compound containing an element other than carbon, hydrogen, or halogen |
| 107 | ...Acetylenic unsaturation containing preservative or stabilizer | 141 | ....Nitrogen containing |
| 108 | ....Hydroxy, bonded directly to carbon, or ether containing | 142 | ....Oxygen containing |
| | | 143 | ...Preparing benzene ring containing compound |
| 109 | ...Nitrogen containing hetero ring in preservative or stabilizer | 144 | ....Haloalkyl containing compound |
| 110 | ...Acyclic nitro containing preservative or stabilizer | 145 | .....By substituting halogen for a different halogen in haloalkyl group |
| 111 | ...Nitrogen other than as ammonia or the ammonium ion in preservative or stabilizer | 146 | ....Forming the benzene ring |
| | | 147 | ....Substituting halogen for different halogen or hydrogen |
| 112 | ....Nitrile | 148 | ...Forming alicyclic ring from benzene ring |
| 113 | ....Imine (e.g., hydrazone, oxime, etc.) | 149 | ...Forming alicyclic ring from acyclic compound |
| 114 | ...Oxygen containing hetero ring in preservative or stabilizer | 150 | ...Preparing from elemental carbon, carbon oxide, or carbon disulfide |
| 115 | ....Hetero ring containing plural ring oxygens | 151 | ...Isomerization |
| | | 152 | ...Decreasing molecular weight of polymer of indeterminate structure |
| 116 | ....Oxirane ring | 153 | ...Preparing unsaturated ompound |
| 117 | ...Carbonyl containing preservative or stabilizer | 154 | ....From acetylenically unsaturated compound |
| 118 | ...Hydroxy, bonded to carbon, or ether containing preservative or stabilizer | 155 | ....By dehalogenation or dehydrohalogenation of adjacent carbon atoms in a compound |
| 119 | ....Phenolic | 156 | .....Catalyst utiliz d |
| 120 | ...Sulfur containing preservative or stabilizer | 157 | ......Alkali or alkaline earth metal containing catalyst |
| 121 | ...Hydrocarbon, halocarbon or halohydrocarbon preservative or stabilizer | 158 | ......Zinc containing catalyst |
| | | 159 | ....From methane or halomethane |
| 122 | ....Acyclic carbon to carbon unsaturation containing | 160 | ....Substituting fluorine for a different halogen |
| 123 | ..Fluorine containing | 161 | ...Utilizing halogen fluoride or a mixture of elemental fluoride and another elemental halogen |
| 124 | ...Product | | |
| 125 | ....Polymer of unsaturated compound | | |
| 126 | .....Fluorine is sole halogen | 162 | ...Utilizing a compound containing silicon and fluorine |
| 127 | ....Benzene ring containing | | |
| 128 | .....Acyclic unsaturation containing | 163 | ...Transhalogenation or disproportionation |
| 129 | .....Plural carbocyclic rings containing | 164 | ...By reacting with hydrogen fluoride |
| 130 | ....Plural carbocyclic rings containing | 165 | ....Catalyst utilized |
| 131 | ....Carbocyclic ring contains six carbon atoms | 166 | .....Metal halide containing catalyst |
| | | 167 | ......Antimony halide containing catalyst |
| 132 | ....Carbocyclic ring contains four carbon atoms | 168 | ......Transition metal halide containing catalyst |
| 133 | ....Carbocyclic ring contains three carbon atoms | 169 | .....Metal oxide containing catalyst |
| 134 | ....Acyclic | | |

*Figure 8.1*   Sample page from the USPTO *Manual of Classification* showing Class 570.

**3. Validity search**: This is the type of search that occurs when it is claimed that some activity infringes another patent, and determines if the claims of that patent are valid; that is, if they define a new invention or if they lack novelty. This involves reviewing the patent file, U.S. and foreign patents in the search area and the literature, to

570-2

CLASS 570    ORGANIC COMPOUNDS -- PART OF THE CLASS 532-570 SERIES

DECEMBER 1996

| | | | |
|---|---|---|---|
| | ORGANIC COMPOUNDS (Class 532, Subclass 1) | 204 | ...Dehalogenation or dehydrohalogenation |
| | | 205 | ....Of alicyclic ring to prepare benzene ring |
| | .HALOGEN CONTAINING | | |
| | ..Fluorine containing | 206 | ...Bonding halogen directly to benzene ring |
| 170 | ...Substituting halogen for a different halogen | 207 | ....Chlorination |
| 171 | ...Increasing the number of carbon atoms in the compound | 208 | .....Catalyst utilized |
| | | 209 | ......Sulfur containing catalyst |
| 172 | ....Utilizing unsaturated compound | 210 | ......Metal halide containing catalyst |
| 173 | ...Decreasing the number of carbons in the compound (e.g., cracking, etc.) | 211 | ...Purification or recovery |
| | | 212 | ..Forming alicyclic ring from benzene ring |
| 174 | ...Introducing bromine or iodine | 213 | ..Purification or recovery of |
| 175 | ...Utilizing unsaturated compound | | 1,2,3,4,5,6 - hexachlorocyclohexane |
| 176 | ...Replacing halogen with hydrogen | | (i.e., benzene hexachloride) |
| 177 | ...Purification or recovery | 214 | ..Ring formation, ring expansion or contraction or bonding one |
| 178 | ....Including distillation | | alicyclic ring directly or |
| 179 | ....Solid sorbent utilized | | indirectly to another alicyclic |
| 180 | ....Including extraction with organic liquid | | ring |
| | | 215 | ...Diels-Alder reaction |
| 181 | ..Product | 216 | ..Processes of preparing, purifying, or |
| 182 | ...Benzene ring containing | | recovering unsaturated compound |
| 183 | ....Polycyclo ring system | 217 | ...From carbon source other than |
| 184 | ....Plural benzene rings bonded directly to the same acyclic carbon or attached by an acyclic carbon chain | | hydrocarbon, halocarbon, or halohydrocarbon |
| | | 218 | ...Decreasing the number of carbon atoms in the compound |
| 185 | ....Benzene ring and halogen bonded directly to the same acyclic carbon or attached by an acyclic carbon chain | 219 | ...Plural diverse reactions in separate zones |
| | | 220 | ....Dehalogenation or dehydrohalogenation with |
| 186 | ...Alicyclic ring containing | | halogenation in separate zones |
| 187 | ....Polycyclo ring system | 221 | .....Acetylene reactant |
| 188 | ....Plural rings containing | 222 | .....Including oxyhalogenation or |
| 189 | ...Acyclic carbon to carbon unsaturation containing | | oxidation with elemental oxygen |
| 190 | ..Processes of preparing, purifying, or recovering benzene ring containing compound | 223 | ....Including oxhalogenation or oxidation with elemental oxygen |
| | | 224 | ...Oxyhalogenation |
| 191 | ...Preparing acyclic haloalkyl group containing compound | 225 | ....Liquid medium or inorganic melt utilized |
| 192 | ....Halo, 1,1-diphenylethane or ring substituted derivative thereof prepared (e.g., DDT, etc.) | 226 | ...Dehydrohalogenation |
| | | 227 | ....Catalyst utilized |
| | | 228 | .....Catalyst in liquid phase |
| 193 | ....Having acyclic carbon to carbon unsaturation | 229 | ....Including chemical reaction with by-product hydrogen halide |
| 194 | ....Bonding haloalkyl group directly to benzene ring | 230 | ...Dehalogenation or dehydrogenation |
| 195 | .....Oxygen containing organic compound reactant | 231 | ...Addition reaction of free halogen or hydrogen halide to carbon to carbon unsaturation |
| 196 | ....Halogenation of acyclic carbon | 232 | ....To triple bond |
| 197 | .....Catalyst utilized | 233 | .....To acetylene |
| 198 | ......Halogen containing catalyst | 234 | ...Elemental halogen reactant |
| 199 | ...Bonding benzene rings to the same acyclic carbon or to an acyclic carbon chain | 235 | ...Metal halide reactant |
| | | 236 | ...Isomerization |
| | | 237 | ...Increasing the number of carbon atoms in the compound |
| 200 | ...Preparing acyclic carbon to carbon unsaturation containing compound | 238 | ...Purification or recovery |
| 201 | ...Oxygen containing organic compound reactant | 239 | ....Including contact with solid agent |
| | | 240 | ..Preparing from elemental carbon, inorganic carbide, carbon disulfide, or carbon oxide |
| 202 | ...Isomerization | | |
| 203 | ...Oxyhalogenation | | |

*Figure 8.2*    Second page of Class 570, from the *Manual of Classification.*

determine if the concept has been disclosed or described before the application date.

Validity and infringement searches are usually required to determine if the patent can be commercially practiced free from legal action. These searches are generally lengthy and cover a broad area.

CLASS 570 ORGANIC COMPOUNDS -- PART OF THE CLASS 532-570 SERIES          570-3

DECEMBER 1996

ORGANIC COMPOUNDS (Class 532, Subclass
   1)
   .HALOGEN CONTAINING
241   ..Preparing utilizing plural diverse
         reactions in separate zones
242   ...Addition reaction of hydrogen
         chloride to carbon to carbon
         unsaturation with chlorination in
         separate zone
243   ..Preparing by oxyhalogenation
244   ...Liquid medium or inorganic melt
         utilized
245   ...Fixed bed catalyst utilized
246   ..Preparing by addition of elemental
         halogen, interhalogen compound, or
         hydrogen halide to carbon to carbon
         unsaturation
247   ...Catalyst or reaction directing agent
         utilized
248   ....Hydrogen halide reactant
249   .....Nonmetallic catalyst or reaction
         directing agent utilized
250   .....Catalyst or reaction directing
         agent containing or group VIII
         metal utilized
251   ...All reactants in vapor phase
252   ..Elemental halogen reactant
253   ...Catalyst or reacton directing agent
         utilized
254   ....Inorganic metal containing catalyst
         or reaction directing agent
         utilized
255   ...All reactants in vapor phase
256   ..Isomerization
257   ..Preparing by increasing the number of
         carbons in the compound
258   ..Preparing by reacting hydrogen halide
         with a compound which contains
         hydroxy bonding directly to carbon
259   ..Preparing by reacting ether with
         hydrogen halide
260   ..Preparing by halogen exchange
261   ..Halogen source is a compound other
         than hydrogen halide
262   ..Purification or recovery
263   ...Liquid-liquid extraction
264   ..Preservation or stabilization
         treatment

*Figure 8.3*  Final page of Class 570, from the *Manual of Classification.*

They are generally performed by a professional searcher, a patent
agent, or an attorney.

4. **Assignment search**: This is a type of search in which one wishes to
   determine the present owner of a patent. This search is usually conducted in the Patent Assignment Branch of the Patent Office and is a
   computer search. The records since 1982 have been placed on a computer; if the patent issued or has been reassigned since that date, one
   can question the computer and obtain the record of the assignments
   from its first recordation. Since a person or corporation who transfers
   ownership of patent rights is required to inform the PTO, this search
   can only tell if the documents have been received and recorded by
   the PTO. The appearance of the name of an assignee on a patent does
   not necessarily mean that the formal assignment has been registered

## INDEX TO CLASSIFICATION

| Half Wave | Class | Subclass |
|---|---|---|
| Vacuum tube type | 363 | 114+ |
| With filter | 363 | 39+ |
| With voltage regulator | 363 | 84+ |
| **Halftone** | | |
| Blanks and processes printing | 101 | 401.1 |
| Etching | 216 | 72 |
| Etching | 216 | 95 |
| Photographic process | 430 | 396 |
| Photographic screens | 359 | 893+ |
| Chemically defined | 430 | 6+ |
| Printing plates | 101 | 935 |
| **Halides (See Material Halogenated)** | | |
| Hydrocarbon | 570 | 101+ |
| As azeotropes | 203 | 67 |
| Electromagnetic wave synthesis | 204 | 157.15+ |
| Electrostatic field or electrical discharge synthesis | 204 | 169 |
| Metal | 423 | 462+ |
| Electrolytic symbols | 205 | 498+ |
| Nitroaromatic | 568 | 927+ |
| Nonmetal inorganic | 423 | 462 |
| Organic acid | 562 | 800+ |
| Rubber hydrohalide | 525 | 332.3 |
| **Hall Effect Means in an Amplifier** | 330 | 6 |
| **Haloamines** | 564 | 114+ |
| Acyclic | 564 | 118+ |
| Hydroxy or ether containing | 564 | 119 |
| Plural difluoramine groups | 564 | 121+ |
| Unsaturated | 564 | 120 |
| Alicyclic | 564 | 117 |
| Amidines | 564 | 116 |
| **Halogen Compounds (See Material Halogenated)** | | |

| | Class | Subclass |
|---|---|---|
| Perforated discharge | 241 | 86+ |
| Process | 241 | 27 |
| Series material flow | 241 | 154 |
| Musical instruments | | |
| Piano | 84 | 236+ |
| Stringed instrument | 84 | 323+ |
| Tuning | 84 | 459 |
| Nut cracker | 30 | 120.1 |
| Pile driver | 173 | 90+ |
| Punching machine | 83 | |
| Riveting | 72 | 476+ |
| Road rammer | 404 | 133.05 |
| Rock drilling | 175 | 135+ |
| Rod encircling type | 81 | 27 |
| Saw stretching machine | 76 | 26 |
| Scale removing | 29 | 81.15 |
| Shoe lasting stretcher and stoneworking combined | 12 | 109 |
| Impact tools | 125 | 6+ |
| Tool driving | 125 | 40 |
| Tube cleaner inside | 173 | 90+ |
| Tuning for pianos | 15 | 104.07 |
| Typewriter | 84 | 459 |
| Bar | 400 | 388+ |
| Key wheel | 400 | 154+ |
| Welding | 228 | 24 |
| Woodworking | 81 | 20+ |
| **Hammock** | 5 | 120+ |
| Berth | 105 | 320 |
| Design | D06 | 386+ |
| Swing | 297 | 277 |
| **Hamper or Basket** | D32 | 37 |
| Laundry | D32 | 37 |

| Handle | Class | Subclass |
|---|---|---|
| Pick for stringed musical instrument | 84 | 322 |
| Sorting | 209 | 614 |
| Telephone dialing tool | 379 | 456 |
| Toaster or broiler | 99 | 394 |
| Muffs design | D02 | 611+ |
| Operated devices (see type of Device operated) | | |
| Operated devices agricultural and Earth working | | |
| Cultivating tools | 172 | 351+ |
| Forks and shovels | 294 | 49+ |
| Harvesting cutter | 56 | 239+ |
| Plant irrigators | 47 | 48.5 |
| Planting dibble | 111 | 99 |
| Planting drill | 111 | 82 |
| Rakes | 56 | 400.01+ |
| Raking and bundling | 56 | 342 |
| Scoop excavator wheeled | 37 | 434 |
| Snow excavator | 37 | 196+ |
| Snow excavator and melter | 37 | 230 |
| Subsoil irrigators | 111 | 7.1+ |
| Organs automatic | 84 | 84+ |
| Setting mechanism for electric clocks | 368 | 60 |
| Stamps | 101 | 405 |
| Design | D18 | 14+ |
| Surgical thermal wear | 607 | 145 |
| Tools (see type of tool) | | |
| Assembling or disassembling | 29 | 270+ |
| Barrel croze | 147 | 24 |
| Barrel head scriber | 147 | 41 |
| Boot and shoe making | 12 | 103+ |
| Butchering intestine cleaners | 452 | 123 |

*Figure 8.4*    Representative page from USPTO *Index to Patent Classification.*

| Entry | Class | Subclass |
|---|---|---|
| **Halogenated Carboxylic Acid Esters** | | |
| Acyclic acid esters | 560 | 1+ |
| Of phenals | 560 | 226+ |
| Acyclic amino acid esters | 560 | 145 |
| Acyclic carbamic acid esters | 560 | 172 |
| Acyclic oxy acid esters | 560 | 161 |
| Acyclic polycarboxylic acid esters | 560 | 184 |
| Acyclic unsaturated acid esters | 560 | 192 |
| Alicyclic | 560 | 219 |
| Aromatic amino acid esters | 560 | 125 |
| Aromatic carbamic acid esters | 560 | 47 |
| Aromatic polycarboxylic acid esters | 560 | 30 |
| Oxybenzoic acid esters | 560 | 23 |
| Phenoxyacetic acid esters | 560 | 65 |
| **Halohydrin** | 568 | 62+ |
| **Halothiocarbonate Esters** | 568 | 841+ |
| **Halowax** | 568 | 249 |
| **Halter** | 570 | 181 |
| Brassiere type garment | 450 | 1+ |
| Feed bags supported on | 119 | 66 |
| Harness | 54 | 24+ |
| Design | D30 | 134+ |
| Poke with bar and | 119 | 758+ |
| Snap releasers | 119 | 776 |
| **Hamburger** | | |
| Cookers | 99 | 422 |
| Grinders | 241 | |
| Molding and shaping | 54 | 25+ |
| Briquetting meat | 54 | 18.1+ |
| **Hames** | | |
| Collar combined | D30 | 137 |
| Design | 54 | 30+ |
| Traces and connectors | 54 | 32+ |
| Tugs | 81 | 20+ |
| **Hammer** | | |
| Automobile fender straightening | 72 | 705* |

| Entry | Class | Subclass |
|---|---|---|
| **Hand** | | |
| Article manipulated by mechanical hand-like movement | 414 | 1+ |
| Artificial | 623 | 57+ |
| Baskets | 217 | 122+ |
| Checker radioactivity | 250 | 336.1 |
| Miscellaneous | 250 | 336.1 |
| Scintillation type | 250 | 361 R+ |
| Clock and watch | 368 | 228 |
| Clock and watch | 968 | 142+ |
| Design | D10 | 127 |
| Coverings | D02 | 610+ |
| Doll | 446 | 327+ |
| Dryers electric | 392 | 380+ |
| Design | D28 | 54.1 |
| Electric applicators | 607 | 145+ |
| Exercising appliances | 601 | 40 |
| Gloves mittens and wristlets | 2 | 158+ |
| Design | D02 | 617+ |
| Grips (see handle) | | |
| Blotter | 34 | 95.2 |
| Bowling ball | 473 | 127+ |
| Coating implement with material supply | 401 | 6 |
| Crutch hand hold | 135 | 72 |
| Golf club | 473 | 300+ |
| Hand wheel rim | 74 | 558 |
| Handle bar | 74 | 551.9 |
| Penholder | 15 | 443 |
| Rein holds | 54 | 74 |
| Ski pole | 280 | 821+ |
| Tennis racket | 473 | 549+ |
| Wrench | 81 | 300+ |
| Guards and protectors | 2 | 16 |
| Cutting hand tool | 30 | 295 |
| Design | D29 | 113+ |

| Entry | Class | Subclass |
|---|---|---|
| Cherry seeder | 30 | 113.1 |
| Cleaning | 15 | |
| Clock or watch | 968 | 651+ |
| Cutlery | 30 | |
| Dental | 433 | 25 |
| Fish scaling | 452 | 105 |
| Forks and shovels | 294 | 49+ |
| Hog scrapers | 452 | 94 |
| Meat and vegetable | 30 | |
| Meat tenderizers | 452 | 141+ |
| Oyster openers | 452 | 17 |
| Peach stoner | 30 | 113.1 |
| Pick, miners | 125 | 43 |
| Plural diverse tools | 7 | |
| Railway mail delivery | 258 | 3 |
| Raisin seeder | 30 | 113.1 |
| Rake | 56 | 400.01+ |
| Tamper for track laying | 104 | 13+ |
| Tufting e.g., chenille applying | 112 | 80.03+ |
| Tying cords or strands | 289 | 17 |
| With light | 362 | 119+ |
| Vehicles | | |
| Land occupant propelled by hand | 280 | 242.1+ |
| Railway hand car | 105 | 86+ |
| Scoop excavator | 37 | 435+ |
| Self loading type | 414 | 444+ |
| Truck ladder | 182 | 127 |
| Truck ladder with erection means | 182 | 63.1+ |
| Gear lifting means | 182 | 69.3 |
| Winch lifting means | 182 | 69.2 |
| Trucks and barrows | 280 | 47.17+ |
| Trucks and barrows dumping type | 298 | 2+ |
| Wheels | 74 | 552+ |
| **Hand Cart Design** | D34 | 12+ |
| **Hand Mirror** | 359 | 882 |
| Plural mirror | 359 | 850+ |

*Figure 8.4 (continued)*

## INDEX TO CLASSIFICATION

| Half Wave | Class | Subclass |
|---|---|---|
| Burglar alarm | 116 | 88+ |
| Claw | 254 | 26 R |
| Combined with additional tools | 7 | 143+ |
| Design | D08 | 75 |
| Drop forging | 72 | 435+ |
| Earth boring tool combined | 175 | 135 |
| Firearm | 42 | |
| Forging | 72 | 476+ |
| Heads for piano actions | 84 | 254 |
| Impact clutch type | 173 | 93.5 |
| Implement combined | 81 | 463+ |
| Awl or prick punch | 30 | 358+ |
| Internal combustion charge igniter | | |
| rocking electrode | 123 | 157 |
| Leather compacting | 69 | 1 |
| Magazine | 227 | 133 |
| Making | | |
| Forging dies for | 72 | 470+ |
| Processes of | 76 | 103 |
| Metal bending | 72 | 462+ |
| Mills | 241 | 185.5+ |
| Parallel material flow | 241 | 138 |

| | Class | Subclass |
|---|---|---|
| Fork | 30 | 323 |
| Machine safety stop | 192 | 130+ |
| Guides rests and straps | | |
| Coffin lowering strap | 27 | 33+ |
| Railway car strap | 105 | 354 |
| Stringed instrument rest | 84 | 328 |
| Typewriter keyboard guide | 400 | 715+ |
| Levers | 74 | 523+ |
| Looms | 139 | 29+ |
| Manipulated implements (see hand Tools) | | |
| Apparel fluting iron | 223 | 36 |
| Atomizer | 239 | 337+ |
| Bars for carrying | 294 | 15+ |
| Bells | 116 | 171 |
| Blowtorch | 431 | 344 |
| Cutlery | 30 | |
| Dental | 433 | 141+ |
| Electric flashlight | 362 | 208 |
| Flatiron | 38 | 95 |
| Lantern | 362 | 257+ |
| Material handling | 294 | |

| Handle | Class | Subclass |
|---|---|---|
| **Handbags** | 150 | 100+ |
| Design | D03 | 232+ |
| Frame | D03 | 324 |
| Latch | D08 | 331+ |
| **Handcuffs** | | |
| Design | 70 | 16+ |
| Carrier | 224 | 914* |
| **Handgun (See Pistol; Revolver)** | | |
| **Handicapped Person Handling** | | |
| Elevator adjacent stairway | 414 | 921* |
| Elevator control | 187 | 200 |
| **Handkerchief** | 187 | 901* |
| Design | D02 | 500+ |
| Garment attaching means | 24 | 3.1+ |
| Garment worn | 2 | 279 |
| Lace | 87 | 10 |
| Medicated | 604 | 358+ |
| **Handle (see Grip)** | 16 | 110 R |
| Article carrier | 294 | |
| Asymmetric-type | | |
| Coffee, teapot and pitcher | D08 | DIG. 9 |
| Lever type | D08 | DIG. 1 |
| Bar | 74 | 555.1+ |

*Figure 8.4 (continued)*

**184.   Plural benzene rings bonded directly to the same acyclic carbon or attached by an acyclic carbon chain:**
Compound under subclass 182 which contains two or more benzene rings bonded directly to the same carbon which is not part of a ring or bonded through two or more carbons, none of which is part of a ring.

**185.   Benzene ring and halogen bonded directly to the same acyclic carbon chain:**
Compound under subclass 182 which contains a benzene ring and a halogen bonded directly to a carbon atom which is not part of a ring or a benzene ring attached to a halogen atom through a chain of two or more carbon atoms, none of which is part of a ring.

**186.   Alicyclic ring containing:**
Compound under subclass 181 which contain a ring of three or more carbon atoms, which ring is not a benzene ring.

**187.   Polycyclo ring system:**
Compound under subclass 186 which contains a ring system of at least two rings which have tow or more carbon atoms in common.

(1)   Note.  These compounds are usually referred to in the art as fused (only two atoms in common) or bridged (three or more atoms in common).

SEARCH CLASS:
552,   Organic Compounds, subclass 653 for halogenated derivatives of Vitamin D compounds, cholecalciferols, activated 7–dehydrocholesterols, dihydrotachys-terols, 3–5 cyclovitamin D compounds, etc.

**188.   Plural rings containing:**
Compound under subclass 186 which contains two or more alicyclic rings.

**189.   Alicyclic carbon to carbon unsaturation containing:**
Compound under subclass 181 which contains carbon to carbon unsaturation which is not part of any ring.

**190.   Processes of preparing, purifying, or recovering benzene ring containing compound:**
Process under subclass 101 wherein a benzene ring containing compound is prepared, purified, or recovered.

**191.   Preparing acyclic haloalkyl group containing compound:**
Process under subclass 190 of preparing a benzene ring containing compound which contains a halogen atom bonded indirectly to a benzene ring through one or more carbon atoms, none of which is part of a ring.

**192.   Halo, 1,1–diphenylethane or ring substituted Derivative thereof prepared DDT, etc.):**
Process under subclass 191 wherein the compound prepared has the 1,1–diphenylethane structure, i.e.,

wherein at least one hydrogen atom is replaced with halogen and in addition may have other substituents on the ring carbon atoms.

**193.   Having acyclic carbon to carbon unsaturation:**
Process under subclass 191 wherein the compound prepared contains carbon to carbon unsaturation which is not part of a ring.

**194.   Bonding haloalkyl group directly to benzene ring:**
Process under subclass 191 wherein an aliphatic carbon atom bonded directly or indirectly to halogen is bonded directly to a benzene ring, or wherein an aliphatic carbon atom is bonded directly to a benzene ring, a halogen is bonded directly or indirectly thereto.

**195.   Oxygen containing organic compound reactant:**
Process under subclass 194 wherein an oxygen containing compound acts as a source of carbon or halogen in the product.

**196.   Halogenation of acyclic carbon:**
Process under subclass 191 wherein halogen is bonded directly to an acyclic carbon atom.

*Figure 8.5*   Representative page from USPTO *Classification Definitions.*

in the U.S. Patent and Trademark Office. Failure to record an assign-ment can bring penalties to the owner of the patent.

5. **Maintenance fee search**: This is a type of search in which one wishes to determine if the fees paid by the owner of the patent have actually been paid and if the patent can still be enforced. With the current

schedule of fee payments, the owner has to pay fees during the 4th, 8th, and 12th years to keep the patent in force. This search is accomplished by checking the database on patent fee payments and determining if the fees have been paid.

Before any of these searches begin, a careful evaluation of the concept, idea, or discovery must be made to clearly define the area of technology that must be investigated. If a search is to be conducted in the U.S. Patent and Trademark Office in Arlington, VA, or if it is to be done using a computerized database, on the Internet, or in a Patent Depository Library, the parameters of the concept must be carefully defined. Computerized searching generally employs the use of keywords to query the database, but how is the keyword determined? As previously stated, the PTO has determined specific areas of technology and divided the area into classes and subclasses. Patents are not written for the "common man;" they are legal documents that define specific legal rights that accrue to the inventor and owner of the patent. The language is not the type used in the scientific literature, and many of the patents are written in terminology that is unique to the specific technical field. The writer of the patent is drafting the documents in as broad a manner as allowed and must, according to law, explain to the patent examiner, how to perform the invention. Some patents are written in language so broad that normal keywords in the art area are never used, and the patent may not be found by keyword searching. Problems that confront the searcher, whether he is computer searching or hand searching, include locating the area where the relevant patents similar to the idea disclosed would be found. Terminology in a subject area is critical. Are window curtains and venetian blinds found in the same class? No! Many classes were defined in the age prior to the adoption of names now in use for the items. Spelling is another problem; consider the term "color." Is it the same as colour? According to the PTO, it is. But if one is using a computer, it is not. In some technical areas, the use of different words to describe a specific part or action of a device will produce confusing results. Acronyms cause other problems. Many scientists use them and assume that all people use the same acronym for the same material; but looking at the polymer patents, one will find that the same acronyms have been used for many different compounds.

How does one conduct a search after the exact area of the problem has been defined? Searches can be conducted in several ways.

- **Hand search:** This is basically a patent-by-patent search in a subclass. This type of search is most easily conducted in the Public Search Room at the Patent Office. Here, the patents are filed by classes and subclasses, so one can find all of the patents in a small section of the files. The patent copies are filed in a set of shelves or trays, which are called "shoes," because early in the history of the Patent Office, shoe boxes were actually used to store the patents. Other locations have collections of patents stored in numerical order, and to search by hand would require one to list the patent numbers in the subclass and then go through the numerical list to find the patents. Chapter 9 will go into more detail.

- **Computer search:** This is basically a search in which one looks for keywords or phrases found in a patent. The database may allow for the searching of the entire patent document or portions thereof (e.g., the title, abstract, claims). The Patent Office has built its own databases, and these are available in the Patent Depository Libraries throughout the U.S. These databases may be limited to certain current patent documents. Many use only the patents issued after 1971, while others go further back. Commercial databases likewise may be limited in scope and range. Computer searching will be discussed in more detail in Chapter 9.

- **Literature search:** This is a search of the published literature to locate patents. This can be easily done in a library that has the Abstract volumes of the particular science concerned. In chemistry, *Chemical Abstracts*, in other fields, *Electronic Abstracts*, *Physical Abstracts*, etc., are typical sources of data. *Chemical Abstracts* also lists in its Patent Section, a cross-reference that gives equivalent patent numbers in foreign countries. This is a valuable tool for locating the countries where patents equivalent to U.S. or foreign patents have been issued. Reference to foreign patents in the literature can also be cross-referenced to see if an equivalent patent exists in the U.S. It should be noted than an equivalent patent might not be an identical patent. The patents all originate from the same parent application; but due to the laws of each country, the prosecution of the application under those laws can lead to differing claims. The essence of the patents will be similar, but the protection gained in each country may be slightly different.

- **Watch searches:** In this procedure, one will return again and again to the search area to determine what has been patented since the last search. In other words, one is constantly updating a prior search and following the progress being made in the area of the invention. This type of search is conducted on a routine basis to keep the investigator abreast of current patented technology. One does not have to go back through the patents again in a subclass, but only has to go to the patents that have issued since the last search. A simple and convenient way to do this is through the *Official Gazette* (*OG*). This publication is published each Tuesday by the U.S. Patent and Trademark Office and is the official list of all patents that were issued on that date. A page of the *OG* has been reproduced in Figure 8.6, and for each patent, the number, the title, the primary class and subclass where the patent is classified, the inventor and assignee, as well as a representative claim, are shown. The patents are listed in sequential order but are assigned that number depending on their primary class and subclass. Also, at the back of each issue, there is a tabulation of classes and subclasses

MAY 11, 1999          CHEMICAL          1369

**5,902,820**
**MICROBICIDAL COMPOSITIONS AND METHODS**
**USING COMBINATIONS OF PROPICONAZOLE WITH**
**DODECYLAMINE OR A DODECYLAMINE SALT**
Percy A. Jacquess, Tigrett; David Oppong; Sheldon M. Ellis, both of Cordova, and L. Fernando Del Corral, Memphis, all of Tenn., assignors to Buckman Laboratories International Inc., Memphis, Tenn.
Filed Mar. 21, 1997, Appl. No. 821,912
Int. Cl.[6] A01N 33/02; 37/30; 43/36; 43/64
U.S. Cl. 514—383         28 Claims
1. A microbicidal composition comprising:
(a) propiconazole and (b) dodecylamine or a dodecylamine salt, wherein (a) and (b) are present in a synergistically effective amount to control the growth of at least one microorganism.

---

**5,902,821**
**USE OF CARBAZOLE COMPOUNDS FOR THE**
**TREATMENT OF CONGESTIVE HEART FAILURE**
Mary Ann Lukas-Laskey, Rosemont; Robert Ruffolo, Jr., Spring City; Neil Shusterman, Wynnewood, all of Pa.; Gisbert Sponer, Laudenbach, and Klaus Strein, Hemsbach, both of Germany, assignors to Boehringer Mannheim Pharmaceuticals Corporation Smith Kline Corporation Limited Partnership No. 1, Gaithersburg, Md.
PCT No. PCT/EP96/00498, § 371 Date Dec. 29, 1997, § 102(e) Date Dec. 29, 1997, PCT Pub. No. WO96/24348, PCT Pub. Date Aug. 15, 1996
PCT Filed Feb. 7, 1996, Appl. No. 875,603
Claims priority, application Germany, Feb. 8, 1995, 195 03 995
Int. Cl.[6] A61K 31/40
U.S. Cl. 514—411        11 Claims
1. A method of decreasing mortality caused by congestive heart failure in a patient in need of such decrease, said method comprising:
administering to said patient first dosages at least daily for a period of from 7 to 28 days, said first dosages each comprising carvedilol,
then administering to said patient second dosages at least daily for a period of from 7 to 28 days, said second dosages each containing carvedilol, and
then administering to said patient third dosages daily for a maintenance period, said third dosages each comprising carvedilol, said third dosages each comprising a daily maintenance dose in the range of from about 10 mg to about 100 mg of carvedilol,
said first dosages each comprising carvedilol in an amount which is 10–30% of said daily maintenance dose,
said second dosages each comprising carvedilol in an amount which is 20–70% of said daily maintenance dose.

---

**5,902,822**
**7-METHYLTHIOOXOMETHYL AND**
**7-METHYLTHIODIOXOMETHYL PACLITAXELS**
Jerzy Golik, Southington, and Dolatrai M. Vyas, Madison, both of Conn., assignors to Bristol-Myers Squibb Company, Princeton, N.J.
Provisional application No. 60/039,480, Feb. 28, 1997. This application Feb. 18, 1998, Appl. No. 25,270.
Int. Cl.[6] A61K 31/335; C07D 305/14
U.S. Cl. 514—449        11 Claims
1. A compound of formula I, or a pharmaceutically acceptable salt thereof

wherein: R is aryl, substituted aryl, $C_{1-6}$ alkyl, $C_{2-6}$ alkenyl, $C_{3-6}$ cycloalkyl, or heteroaryl;
$R^A$ is hydrogen;
$R^B$ is independently —NHC(O)-aryl, —NHC(O)-substituted aryl, —NHC(O)-heteroaryl, —NHC(O)OCH$_2$Ph, or —NHC(O)O—($C_{1-6}$ alkyl);
$R^C$ is hydrogen;
$R^D$ is hydroxy;
$R^2$ is phenyl or substituted phenyl;
$R^4$ is methyl, ethyl, propyl, cyclopropyl or —O—($C_1$-$C_3$ alkyl);
L is O;
$R^6$ and $R^{6'}$ are hydrogen;
One of $R^{7'}$ and $R^7$ is hydrogen and the other is —OCH$_2$S(O)$_n$CH$_3$;
n=1 or 2;
$R^9$ and $R^{9'}$ are independently hydrogen or hydroxy or $R^9$ and $R^{9'}$ together form an oxo (keto) group;
$R^{10}$ is hydrogen, hydroxy or —OC(O)—($C_1$-$C_6$ alkyl);
$R^{10'}$ hydrogen;
$R^{14}$ is hydrogen or hydroxy; and
$R^{19}$ is methyl.

---

**5,902,823**
**METHOD FOR TREATING ADDICTION USING**
**FORSKOLIN OR EXTRACTS CONTAINING FORSKOLIN**
Paolo Morazzoni, and Ezio Bombardelli, both of Milan, Italy, assignors to Indena S.p.A., Milan, Italy
PCT No. PCT/EP96/01952, § 371 Date Nov. 17, 1997, § 102(e) Date Nov. 17, 1997, PCT Pub. No. WO96/36332, PCT Pub. Date Nov. 21, 1996
PCT Filed May 9, 1996, Appl. No. 952,472
Claims priority, application Italy, May 19, 1995, MI95A1023
Int. Cl.[6] A61K 31/35; 35/78
U.S. Cl. 514—453        5 Claims
1. A method for the treatment of a patient suffering from alcohol addiction, comprising administering to said patient a pharmaceutical composition comprising a therapeutically effective amount of forskolin or an extract containing forskolin in a pharmaceutically acceptable carrier.

---

**5,902,824**
**PHENYLDIHYDROBENZOFURANES**
Wolf-Rüdiger Ulrich, Konstanz, Germany, assignor to Byk Gulden Lomberg Chemische Fabrik GmbH, Konstanz, Germany
PCT No. PCT/EP96/02031, § 371 Date Nov. 18, 1997, § 102(e) Date Nov. 18, 1997, PCT Pub. No. WO96/36625, PCT Pub. Date Nov. 21, 1996
PCT Filed May 11, 1996, Appl. No. 952,276
Claims priority, application Switzerland, May 18, 1995, 1472/95
Int. Cl.[6] A61K 31/34; C07D 307/94; 307/86
U.S. Cl. 514—462        12 Claims
1. A compound of formula I

*Figure 8.6* Typical page from the *Official Gazette (O.G.)*, a weekly publication of the U.S. Patent and Trademark Office.

and a listing of the issued patents that fall into that category. The *OG* makes enjoyable reading for someone interested in patents, as it shows what areas of technology are active. It will also be seen from each issue how many patents were issued that week.

Whatever the method of searching, the primary aim is to be complete. Take notes and make copies of what is found. If it appears that the concept has not been patented, and a patent application is to be filed, an Information Disclosure Statement must be filed in the PTO by the inventor and his attorney or agent. This document accompanies the patent application, or is submitted shortly after filing, and is required by law and discussed in Chapter 6. This statement requires the presentation of all known facts and summary of knowledge to the PTO concerning the pertinent area of invention.

Searching, whatever the method or combination of methods, becomes a boring and tedious operation. The most common error is the inability of the searcher to keep records of where he has searched and what he has searched, even if he has found nothing pertinent. Professional searchers do not always find the search area in the first class or subclass they search; it may take considerable time to locate the right area. If one is searching at the Patent Office in Arlington, the searcher can consult with an examiner who works in the area of the invention to ascertain the classes and subclasses where he or she should look to locate relevant art.

Searches do not have to be conducted in the U.S. Patent Office. It may be more convenient, easier, and faster to conduct the search there, but if one does not live in the area, there are other locations for searching. The Patent Depository Library (PDL) system, discussed in Chapter 7, places copies of U.S. patents in libraries throughout the U.S., which are available for use at no charge. In addition, each of these libraries has aids to assist in the search. These aids consist of computer systems called CASSIS (Classification And Search Support Information Systems). This system contains the *Manual of Classification, Index to Patent Classifications, Definitions*, and other aids. These aids are supplied by the PTO and are updated regularly. In addition, the staff in these libraries have received training to assist the visitor in the search process. These libraries do not all have the same collection of patents, but each will have a collection that will enable one to perform a meaningful search. These libraries will have the patents bound in volumes, on microfilm, or on CD-ROMs. Copies of the patents can be made for later reference.

One feature of the CASSIS databases that is most useful, if one is in the Patent Office Search Room or in a PDL facility, is the use of the database to determine the present location of a patent. This allows one, knowing only the number to query the database, to determine where the patent is presently classified and where it is cross-indexed. As stated earlier, the classification system is in a constant state of flux; areas are being redefined, narrowed, or broadened. As a result, patents are changed from one subclass to another.

This database allows one to find their present position. This may be useful for locating a starting point for a search.

If an item cannot be found in the index that can lead to a search area, ask for help. If the search is being conducted in the Patent Office, help can be obtained from the PTO personnel assigned to the Search Room, as they are specially trained to help searchers. If one is in a PDL facility, ask the clerk for help; he/she has also received training. At the Patent Office, one can also inquire of the patent examiners for specific assistance in locating the search area. They will not assist in the search but can advise as to the proper areas; however, one must approach an examiner assigned to the technology area of the search.

A problem that confronts all searchers is defining the search area. Examples are given below to help the searcher in defining the search area.

## Example 1

One can use an issued patent and define the search area from that. U.S. Patent 5,661,863, shown in Figures 8.7, 8.8, and 8.9, is titled a "Top Removing Tool." The drawings shown in Figure 8.8 show the device and its usage.

How would this gadget be described to locate an area to conduct a search? Of course, on the title page, the patent classification and the examiner's search area are listed; but ignoring that for the moment, what keywords or phrases could define this device in the *Index to the Classification System*? Using the abstract on the title page, one learns that the device is a "tool for opening crimped overlying lids." Looking at the first paragraph of the specification (Figure 8.9), it states that the tool is for opening cans, particularly larger cans such as 3- or 5-gallon or larger cans in which liquid material is stored (e.g., large paint cans). One can generate a list of keywords or phrases to search in the *Index to the Classifications*; one might use Can Openers? Closures? Covers? Tops? Lids? Crimped Lids?

Looking in the *Index*, one finds Can Opener, Class 30, Subclass 400+. (The + sign means that subclasses starting with 400 fall into this area.) Also under Can Openers, one finds Class 7, Subclass 151+ as well as Class 220, Subclass 260. Closures, Covers, and Tops are all cross-referenced under Covers. Lids and Crimped Lids are not listed. These three classes — 30, 7, and 220 — should cover the area and, through cross-referenced classes, lead to the proper area to be searched.

Turning to the *Manual*, one finds Class 30, Subclass 400, titled Can Opener; and going down the subclasses, one finds Subclass 443 "including punch." Class 7 is titled "Compound Tools," and Subclass 156 is titled "Receptacle Opener or Closure Remover, with cutter." Class 220 is titled "Receptacles," and Subclass 260 is titled "with closure opening arrangements for means (e.g., opening devices)." Searching each of these subclasses, one should find gadgets similar to the item and would have noticed that many of the patents are cross-referenced to the other classes.

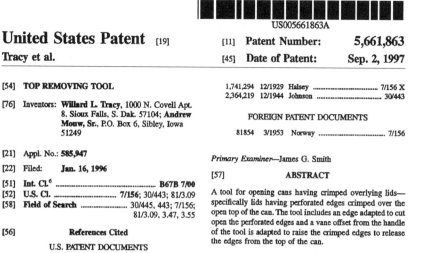

US005661863A

## United States Patent [19]

Tracy et al.

[11] Patent Number: **5,661,863**

[45] Date of Patent: **Sep. 2, 1997**

[54] **TOP REMOVING TOOL**

[76] Inventors: **Willard L. Tracy**, 1000 N. Covell Apt. 8. Sioux Falls, S. Dak. 57104; **Andrew Mouw, Sr.**, P.O. Box 6, Sibley, Iowa 51249

[21] Appl. No.: **585,947**

[22] Filed: **Jan. 16, 1996**

[51] Int. Cl.⁶ .................................................. B67B 7/00
[52] U.S. Cl. .............................. 7/156; 30/443; 81/3.09
[58] Field of Search ...................... 30/445, 443; 7/156; 81/3.09, 3.47, 3.55

[56] **References Cited**

U.S. PATENT DOCUMENTS

1,739,457   12/1929   Harrison ............................ 7/156 X

1,741,294   12/1929   Halsey ............................... 7/156 X
2,364,219   12/1944   Johnson ............................... 30/443

FOREIGN PATENT DOCUMENTS

81854   3/1953   Norway ............................... 7/156

*Primary Examiner*—James G. Smith

[57]                     **ABSTRACT**

A tool for opening cans having crimped overlying lids— specifically lids having perforated edges crimped over the open top of the can. The tool includes an edge adapted to cut open the perforated edges and a vane offset from the handle of the tool is adapted to raise the crimped edges to release the edges from the top of the can.

3 Claims, 1 Drawing Sheet

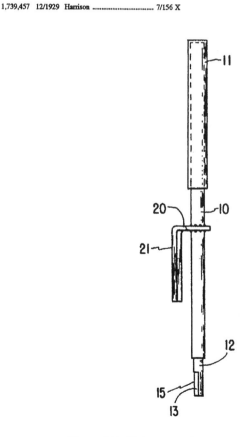

*Figure 8.7*   Title page of U.S. Patent 5,661,863.

U.S. Patent            Sep. 2, 1997              5,661,863

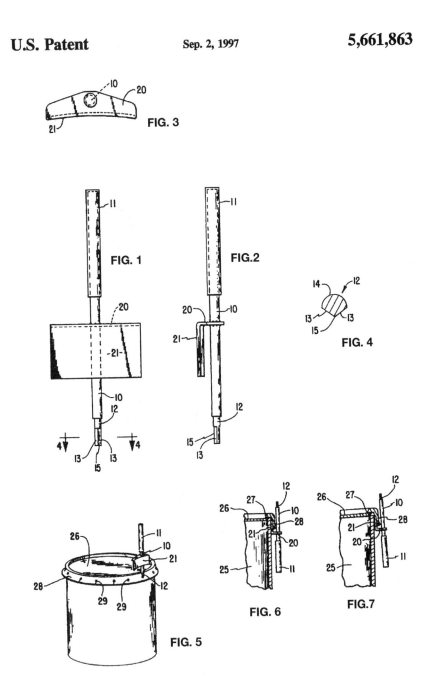

*Figure 8.8*   Page of drawings for U.S. Patent 5,661,863.

5,661,863

**1**

### TOP REMOVING TOOL

#### BACKGROUND AND SUMMARY OF THE INVENTION

This invention pertains to tools for the opening of cans, particularly the larger cans such as 3-gallon, 5-gallon or larger cans in which liquid material is stored.

Most cans for liquid storage holding more than about a gallon or a gallon and a half are now closed by lids having peripheral edges crimped tightly over the upper edge of the can. These edges are often perforated by a series of holes spaced all around the periphery.

Opening these cans requires considerable force in raising the crimped edge of a fairly heavy metal lid. The initial sealing crimp may be applied by machines exerting substantial force. However, in the field, any machine for opening the can is frequently unavailable. Therefore, some other expedient is desirable.

Cans having tops with a continuous rim are particularly difficult to open because the stiffness of a continuous rim far exceeds that of a segmented rim. Thus, there is an obvious need for some means of leverage for opening such cans.

Present expedients may include screwdrivers, wrecking bars or the like to pry under the rim and use such leverage as may be available to spread the crimped edge of the lid. Such devices tend to slip from under the crimped edge and therefore to be inefficient in their operation.

By the present invention, a tool is provided which combines two functions. First it provides a cutting edge to cut the lid into a series of segments, thus resulting in a far less stiff edge to be spread. Second, a convenient lever is provided to pry open the segments or even an unsegmented lid.

#### BRIEF DESCRIPTION OF THE DRAWINGS

FIG. 1 is a plan view of the tool,

FIG. 2 is an edge view of the tool of FIG. 1,

FIG. 3 is an end view from the handle end of the tool of FIG. 1,

FIG. 4 is a sectional view from line 4—4 of FIG. 1,

FIG. 5 is a view of the tool in use on a container for liquid,

FIG. 6 is a detailed view showing the container section and the placement of the tool on the lid of the container, and

FIG. 7 is a view similar to FIG. 6 in a somewhat heavier container.

#### DESCRIPTION

Briefly, this invention comprises a tool having a cutting edge adapted to cut the rim of a container top into segments and also to provide substantial leverage to the edge of the crimped top to open the crimping to allow removal of the top.

More specifically, and referring to the drawings, the tool comprises a body 10. This body may be in the form of a bar or a cylinder of considerable length in order to provide adequate leverage for the required purpose. A handle 11 may be provided as part of the body 10. The handle may be covered with a resilient material, if desired, or may simply be of a larger diameter than the rest of the body to provide less concentrated pressure on the hand of the user. Other forms of handle could also be used. For example, applicants have also used a handle formed of the same material as the body 10, but bent into a triangle.

At the end of the tool opposite the handle 11 is a hardened tip 12, having a cross section as shown in FIG. 4. As shown

**2**

there, the cross section is basically circular except that a pair of flat sides 13 are cut from the arcuate periphery 14. In this way, a sharp edge 15 can be formed from the hardened metal of the tip.

On the shank of the body 10 a top-engaging flange is formed to provide easy engagement to open the crimping of the top. The flange blade 21 formed in arcuate shape as shown (FIG. 3) and having one end bent to form a right angle to the flange. The angular member 20 formed can encircle the body 10 and be fastened to the body by welding or the like to hold the blade 21 spaced from but substantially parallel to the body 10 as shown in FIGS. 2, 6 and 7.

The use of the device is shown in FIGS. 5—7 of the drawing where it is shown in connection with a can 25. The top 26 is attached to the upper edge 27 of the can walls by means of a crimped rim 28 on the top 26. This rim 28, in present usage, often has a series of holes 29 punctured in the rim. Because the present tool is particularly useful in use with this type of rim, it is the one illustrated in FIG. 5. In that figure, the use of the tool at the start of the opening process is shown. The tip 12 of the tool is inserted into a hole 29 with the edge 15 facing the outer periphery of the rim 28. Then, as the handle 11 of this tool is forced toward the center of the top 26, there is considerable leveraged force on the edge of this hole 29. This tends to provide substantial force in a direction to do two things: a) to force the crimped rim 28 in a direction to relieve the crimping, and b) to cut through the metal of the rim 28 by means of the sharpened edge 15. The latter force may be the more desirable. If the edge 15 can be forced through the metal around the hole 29, the rim 28 will be cut into segments which can more readily be bent away from the can. Bent in either case, the rim 28 will tend to be moved to release the pressure that holds the top 26 onto can 25.

The blade 21 may also be used for the purpose of relieving that pressure. Particularly where the edge 15 does cut the rim into segments, but also where the rim is left whole, the blade 21 can be inserted under the flanges of the rim 28, and the handle 11 pulled up and away from the can 25. Again, considerable leverage is available to increase the force available to pull the rim 28 away from the can 25 and to release the crimping pressure.

Thus, the present invention provides a relatively convenient tool for removing can lids which heretofore have caused minor problems for many people.

I claim as my invention:

1. A tool for removing a lid having a rim crimped onto a can, said rim being formed with a series of holes therein, said tool comprising a bar member having a handle and a tip; said tip having a substantially cylindrical cross section with a longitudinally extending cutting edge; said tip adapted to be inserted into any of said holes and then moved over said lid to cut through said holes as said tool is pivoted toward a center of said lid; an arcuate flange secured intermediate said handle and said tip whereby said flange engages an underside of said rim to remove said lid by a prying action.

2. The tool of claim 1 in which said flange includes a blade held in spaced relation to said bar member, said blade being engageable with said underside of said rim.

3. The tool of claim 2 in which said blade includes an angular member at approximately 90 degrees to said blade, said angular member being fixed to said bar member whereby said spaced relation is maintained.

\* \* \* \* \*

*Figure 8.9*   Page of specifications and claims for U.S. Patent 5,661,863.

Looking at the title page of the patent, one sees in the examiner's search area [58] that he searched Class 30/445, 443; Class 7/156; and Class 81/3.09, 3.47, 5.55. He classified the patent [52] in Class 7/156, 30/443, and 81/3.09, with the primary area (in the boldface type) in Class 7, Subclass 156. Class 81 was not indicated in the index and is titled "Tools." One finds Subclass 3.07 "Receptacle Closure Remover" and Subclass 3.09 modifying this subclass titled "Receptacle Closure Remover, combined or plural." Searching Class 30, Subclass 443, one will find many references to other patents all in Class 81, Subclass 3.09, which would indicate searching in that location.

## Example 2

In this example, one is interested in searching a chemical reaction, that of adding chlorine to an aromatic compound resulting in placing a chlorine atom onto the ring. Using the *Index* to look for "chlorinating," one finds an entry, Class 570. By looking in the *Manual* at this class, one will find "Organic Compounds, Halogen Containing" (see Figures 8.1, 8.2, and 8.3). One also finds more than 150 subclasses that subdivide this area of patents. Notice that this class is not limited to chlorination but encompasses all halogens. One should be able to find a subclass that will describe the particular reaction being sought and lead to the proper search area. In chemical reactions, it is also very important to look at the end product. The end product may define the mechanism of its preparation or formation and determine the search area.

Look at the specific patent, U.S. Patent 5,852,222, shown in Figures 8.10, 8.11, and 8.12. This patent is concerned with a method of fluorinating a tetrachlorobutane with hydrogen fluoride in a liquid phase. Broadly speaking, one is fluorinating a halogenated organic compound, and thus is still in Class 570, shown in Figures 8.1 to 8.3. If one looks at this Class, one will reach Subclass 123, "Fluorine Containing." Proceeding down the chart, one next finds "Product," and still farther "Acyclic," Subclass 134. One has probably found the starting point for the search. Looking at item [58] on the title page, one finds that the examiner searched 570/134, and also 169. Subclass 169 is still a division of "Fluorine Containing Products" and adds the provision, "Metal oxide containing catalyst," which is employed in the process.

Always make a quick survey of the patents in the subclass selected to see if anything looks similar, as this may not be the proper subclass. Professional searchers do not always select the right subclass the first time. The definitions used by the Patent Office may not agree with one's selection of the subclass, so it may take searching through several subclasses to locate the proper area.

Unless one has defined the area of the search into a very narrow area and the Patent Office defined the area in the same manner, it will usually be required to search more than one subclass. In order to be complete and thorough, several subclasses will usually be necessary to complete the search. At times, with a complicated device or reaction, a search may cover 10 to 15 subclasses, and 3 or more classes.

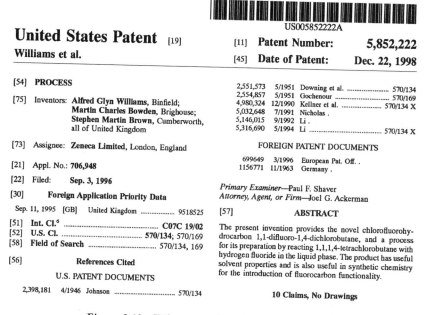

United States Patent [19]

Williams et al.

[11]  Patent Number:     5,852,222

[45]  Date of Patent:     Dec. 22, 1998

US005852222A

[54]  PROCESS

[75]  Inventors: **Alfred Glyn Williams**, Binfield;
**Martin Charles Bowden**, Brighouse;
**Stephen Martin Brown**, Cumberworth,
all of United Kingdom

[73]  Assignee:  **Zeneca Limited**, London, England

[21]  Appl. No.: **706,948**

[22]  Filed:    **Sep. 3, 1996**

[30]      **Foreign Application Priority Data**

Sep. 11, 1995  [GB]   United Kingdom .................. 9518525

[51]  Int. Cl.⁶ .................................... C07C 19/02
[52]  U.S. Cl. .................... **570/134**; 570/169
[58]  Field of Search ................ 570/134, 169

[56]            **References Cited**

U.S. PATENT DOCUMENTS

2,398,181  4/1946  Johnson ........................... 570/134

2,551,573  5/1951  Downing et al. ...................... 570/134
2,554,857  5/1951  Gochenour ........................... 570/169
4,980,324  12/1990  Kellner et al. ................... 570/134 X
5,032,648  7/1991  Nicholas .
5,146,015  9/1992  Li .
5,316,690  5/1994  Li ............................... 570/134 X

FOREIGN PATENT DOCUMENTS

699649   3/1996   European Pat. Off. .
1156771  11/1963  Germany .

*Primary Examiner*—Paul F. Shaver
*Attorney, Agent, or Firm*—Joel G. Ackerman

[57]            **ABSTRACT**

The present invention provides the novel chlorofluorohy-drocarbon 1,1-difluoro-1,4-dichlorobutane, and a process for its preparation by reacting 1,1,1,4-tetrachlorobutane with hydrogen fluoride in the liquid phase. The product has useful solvent properties and is also useful in synthetic chemistry for the introduction of fluorocarbon functionality.

**10 Claims, No Drawings**

*Figure 8.10*  Title page of U.S. Patent 5,852,222.

After finding the search area, do not forget to keep notes as to classes and subclasses that have been covered. Even after finding nothing pertinent, make a note as to what was in that area, so at a later date, when reviewing to determine if the search was thorough and complete, one will know what has been covered. If it is believed that an adequate search has been conducted and no patent has been found that discloses the concept, it may be a winner and a patentable idea! STOP. Do not be overly optimistic; review the definition of the invention again to be sure.

5,852,222

## 1
## PROCESS

The present invention relates a novel chlorofluorohydro-carbon and to a process for its preparation. More particularly it relates to 1,1-difluoro-1,4-dichlorobutane and a process for preparing it from the known compound 1,1,1,4-tetrachlorobutane.

Accordingly the present invention provides 1,1-difluoro-1,4-dichlorobutane. In a further aspect the present invention provides a process for preparing 1,1-difluoro-1,4-dichlorobutane comprising reacting 1,1,1,4-tetrachlorobutane with hydrogen fluoride in the liquid phase.

The process of the present invention is illustrated by the following reaction scheme:

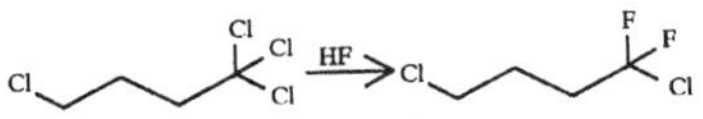

The reaction is conveniently conducted in a vessel whose lining is resistant to corrosion by chemical reaction with hydrogen fluoride, such as for example, one made from "Hastalloy" (Registered Trade Mark) or Monel metal.

The reaction can conveniently be carried out in the presence of a catalyst such as a polyvalent metal halide. Examples of suitable catalysts include ferric chloride, particularly in the presence of activated charcoal, aluminium fluoride, aluminium oxide (γ-alumina), chromium fluoride, manganese difluoride, ferric fluoride, cobalt dichloride, nickel difluoride, zirconium fluoride, thorium fluoride, oxy-fluorides and antimony pentachloride, particularly in the presence of activated charcoal.

Tin halides are preferred catalysts and a particularly useful catalyst is tin (IV) chloride.

The reaction temperature is preferably within the range 50° to 10° C., and more preferably within the range 70° to 90° C. The duration of the reaction is usually within the range 4 to 10 hours.

The reaction is carried out using hydrogen fluoride which is a volatile material having a boiling point under normal atmospheric pressure of 19.5° C. In order to conduct the reaction in the liquid phase a sealed reaction vessel may be used in which the reaction proceeds under the autogenic pressure of the reactants and products. In a preferred variant of this process a vessel can be used which is equipped with means to permit the hydrogen chloride produced during the reaction to to vented, preferably continuously, whilst the reaction is maintained in the liquid phase by the autogenic pressure of the reactants and products. This may be achieved by the use of a condenser which liquifies evaporating hydrogen fluoride whilst permitting the escape of the more volatile hydrogen chloride gas. Such an arrangement permits the autogenic pressure to be maintained in the range of about 175 to about 230 psig (about 12 to about 16 bar).

The product mixture consists principally of the desired 1,1-difluoro-1,4-dichlorobutane, with minor quanties of other materials present, particularly 1,1,1-trifluoro-4-chlorobutane. When the reaction is conducted under a temperature of 85° to 90° C. with venting of the hydrogen chloride over a 6 to 7 hour period good yields and conversion rates may be obtained with minimal co-production of the 1,1,1-trifluoro-4-chlorobutane. Isolation of the desired product can be achieved readily by fractional distillation.

1,1-difluoro-1,4,-dichlorobutane is a novel compound which has useful properties as a solvent, and may be used, for example, in degreasing electrical and electronic components such as printed circuits and the like. Because of its

## 2

higher boiling point and lower volatility compared with the halomethanes and haloethanes traditionally used for degreasing, and the fact that it is a chlorofluorohydrocarbon and not a chlorofluorocarbon, its use may have environmental advantages. It is also of use as a synthetic chemical intermediate particularly for introducing fluorocarbon functionality into a molecule, for example as a means of introducing the difluorobutenyl group into the nematicidal pyrimidine compounds of International Patent Application no. PCT/GB 93/01912.

Various further preferred features and embodiments of the present invention will now be described with reference to the following non-limiting examples. The following abbreviations are used: NMR=nuclear magnetic resonance; s=singlet; d=doublet; dd=double doublet; t=triplet; q=quartet; m=multiplet; br=broad; M=mole; mM=millimoles; $CDCl_3$=deuteriochloroform. Chemical shifts (δ) are measured in parts per million from tetramethylsilane. $CDCl_3$ was used as solvent for NMR spectra unless otherwise stated.

### EXAMPLE 1

5 g 1,1,1,4-Tetrachlorobutane (25 mmoles) was charged to a 25 ml Monel autoclave, which was then purged. Hydrogen fluoride 10.6 g (535 mmoles) was added as liquified gas, the stirrer started and the vessel heated to 80° C. at a ramp rate of 1 deg/min where it was stirred for 18 hours by which time the pressure had increased to 298 psi. The heating was turned off to allow the reaction to cool to room temperature. After the temperature had dropped to ca. 20° C. the vessel was cooled in an ice/IMS bath and the excess pressure (154 psi at room temperature) vented via a stirred water trap keeping the internal temperature >0° C. to reduce the loss of entrained volatile products. On completion of the venting the vessel was opened and the dark red reaction mixture was poured carefully onto ice (ca. 50 gms), the organic phase separated, small amounts of sodium fluoride and magnesium sulphate were added to absorb any hydrogen fluoride and water. The weight of this liquid before the addition of the NaF/$MgSO_4$ was 1.7 gms. The aqueous liquors were extracted with dichlorobenzene (2×30 mls) and the extracts backwashed with water and dried over magnesium sulphate.

Analysis: Analysis by GC (gas chromatography) of the recovered 1.7 g of sample indicated: 0% starting material, 11% 1-fluoro-1,1,4-trichlorobutane, 57% 1,1-difluoro-1,4-dichlorobutane (desired product).

$^1$Hnmr ($CDCl_3$): 2.15 (m, 2H, $CH_2$); 2.50 (m, 2H, $CH_2CF_2Cl$); 3.55 (br t, 2H, $CH_2Cl$).

MS: 142 (M$^+$-HF), 127 (M$^+$-Cl).

### EXAMPLE 2

5.5 g 1,1,4-Tetrachlorobutane (28 mmoles) was charged to a 25 ml Monel autoclave, which was then purged. Hydrogen fluoride 10.1 g (505 mmoles) was added as a liquified gas the stirrer started and the vessel heated to 30° C. at a ramp rate of 1 deg/min. The initial pressure at this temperature was 27 psi, this rose to 36 psi while the reaction was stirred overnight. This rate of pressure increase was not considered to be sufficient so the reaction temperature was increased to 50° C. and the reaction stirred for a further 23 hours while the pressure increased from 47 psi to 106 psi. The vessel was cooled in an ice/IMS bath and the excess pressure (72 psi at room temperature) vented via a stirred water trap keeping the internal temperature <0° C. to reduce the loss of entrained volatile products. On completion of the

*Figure 8.11*    Page of specifications and examples for U.S. Patent 5,852,222.

5,852,222

3

venting the vessel was opened and the dark red reaction mixture was poured carefully onto ice (ca. 50 gms) and the organic phase separated, small amounts of sodium fluoride and magnesium sulphate were added to the straw coloured liquid to absorb any hydrogen fluoride and water. The damp weight of the material was 2.85 g. The aqueous liquors were extracted with with dichlorobenzene (2×30 mls) and the extracts backwashed with water and dried over magnesium sulphate. GC analysis indicated the presence 1,1-difluoro-1,4-dichlorobutane.

EXAMPLE 3

4.9 g 1,1,1,4-Tetrachlorobutane (25 mmoles) was charged to a 25 ml Monel autoclave, which was then purged. Hydrogen fluoride 10.7 g (535 mmoles) was added as a liquified gas, the stirrer started and the vessel heated to 65° C. at a ramp rate of 1 deg/min. The initial pressure at this temperature was ca. 70 psi, this rose to 184 psi over the next 23 hours. After allowing the temperature to drop to ca. 20° C. the vessel was cooled in an ice/IMS bath and the excess pressure (120 psi at room temperature) vented via a stirred water trap (no indication of carry over into this trap) keeping the internal temperature <0° C. to reduce the loss of entrained volatile products (the weight of the vessel dropped by approx. 1 gm during this process). On completion of the venting the vessel was opened and the dark red reaction mixture was poured carefully onto ice (ca 50 gms) and the organic phase separated, small amounts of sodium fluoride and magnesium sulphate were added to the straw coloured liquid to absorb any hydrogen fluoride and water. Damp weight of material was ca. 1 gm. The aqueous liquors were extracted with dichlorobenzene (2×30 mls) and the extracts backwashed with water and dried over magnesium sulphate. GC analysis indicated the presence of the desired product, 1,1-difluoro- 1,4-dichlorobutane.

EXAMPLE 4

2.0 g 1,1,1,4-tetrachlorobutane (10 mmoles) was charged to a 25 ml Monel autoclave, which was then purged. Hydrogen fluoride 9.8 g (490 mmoles) was added as a liquified gas, the stirrer started and the vessel heated to 80° C. at a ramp rate of 1 deg/min. The initial pressure at this temperature was 113 psi, this rose to 161 psi over the next 2 hours 20 minutes before the reaction was left to stir overnight, still at 80° C. The heating was discontinued and the reaction allowed to cool to room temperature. The vessel was cooled in an ice/IMS bath and the excess pressure (78 psi at room temperature) vented via a caustic scrubber keeping the internal temperature <0° C. to reduce the loss of entrained volatile products. On completion of the venting the vessel was opened and the dark red reaction mixture was poured carefully onto ice (ca. 50 gms) and the organic phase extracted into dichloromethane (3×15 mls). The extracts were analysed by GC which suggested that there were two major products (>5% level) with no starting material left. The extracts were dried over magnesium sulphate and the dichloromethane distilled off at atmospheric pressure to give 1.76 g of a dark liquid.

4

GC analysis indicated that the recovered sample contained 36% of the desired product, 1,1-difluoro-1,4-dichlorobutane.

EXAMPLE 5

This Example illustrates the preparation of 1,1-difluoro-1,4-dichlorobutane in the presence of tin (IV) chloride.

1,1,1,4-Tetrachlorobutane (35.3 g), liquified hydrogen fluoride (20.5 g) and tin(IV) chloride (2.6 ml) were charged sequentially at −20° C. into a Monel autoclave fitted with a metal condenser cooled to −15° C. topped with a needle valve to permit venting of gases. The autoclave temperature was raised to 90° C. at ramp rate of 2° C. and maintained at this temperature for 4 hours with periodic venting of the hydrogen chloride produced so as to maintain the internal pressure within the range 180 to 220 psi. The autoclave was then cooled to −10° C. and the contents added carefully to ice (50 g). After allowing the ice to melt the mixture was extracted with dichloromethane (2×20 ml), the extracts combined and dried over sodium fluoride and magnesium sulphate, and the product mixture recovered by evaporation of solvent. Gas chromatographic analysis indicated the presence of a mixture of ca. 79% of the desired 1,1-difluoro-1,4-dichlorobutane and 18% of 1,1,1-trifluoro-4-chlorobutane. The 1,1-difluoro-1,4-dichlorobutane was separated by fractional distillation and obtained as a colourless liquid (20.74 g, b.p 63°–65° C. at 138 mbar).

We claim:

1. 1,1-Difluoro-1,4-dichlorobutane.

2. A process for preparing 1,1-difluoro-1,4-dichlorobutane comprising reacting 1,1,1,4-tetrachlorobutane with hydrogen fluoride in the liquid phase under autogenic pressure.

3. A process according to claim 2 carried out in the presence of a catalyst selected from polyvalent metal halides and aluminium oxides.

4. A process according to claim 3 wherein the metal halide is selected from ferric chloride, aluminium fluoride, chromium fluoride, manganese difluoride, ferric fluoride, cobalt dichloride, nickel difluoride, zirconium fluoride, thorium fluoride, oxyfluorides and antimony pentachloride, optionally in the presence of activated charcoal.

5. A process according to claim 2 wherein the metal halide is selected from halides.

6. A process according to claim 5 wherein the tin halide is tin(IV) chloride.

7. A process according to claim 2 carried out at a temperature within the range 50° to 100° C.

8. A process according to claim 2 carried out under autogenic pressure in a closed vessel.

9. A process according to claim 2 carried out under autogenic pressure in a vessel permitting continuous venting of hydrogen chloride gas produced by the reaction.

10. A process according to claim 9 in which the autogenic pressure is maintained within the range of about 175 to about 230 psig (about 12 to about 16 bar).

* * * * *

*Figure 8.12* Page of examples and allowed claims for U.S. Patent 5,852,222.

# chapter nine

# Searching by hand or computer?

Having entered the "computer age" and since the Internet has become so popular, many believe that the use of a computer is the answer to all problems. But in the field of patent searching, this is not necessarily true. Chapter 8 described the types of searching required to review the patents issued by the U.S. Government to answer the question sought. In broad terms, today's method of searching can be broken into two distinct types: working by hand or through the use of a computer on a database of U.S. patents.

Searching was done only by hand until very recent times. It was known that new methods of searching had to be found as the pool of patent documents was rapidly growing. In the 1930s, ideas were proposed to "automate searching." Methods involving punched cards, microfilm, and microfiche images on individual cards were proposed and studied. Research was undertaken by many of the patent agencies throughout the world, but nothing was successful until the modern computer arrived. Industrial companies having mainframe computers in-house, placed their patents on the computer and created a database with which they could experiment. Many ideas resulted in the creation of databases and methods of searching. But it was not until the early 1990s that the abilities of the computer and the programmer could produce a system that was workable. As time progressed, the systems improved, the databases grew larger, electronic scanning was perfected, and the ability to produce the patent document on the computer screen, identical to the printed hard copy, could be demonstrated and put into use.

## Searching by hand

For hand searching, it is preferable to conduct the search in the Public Search Room of the U.S. Patent and Trademark Office, in Crystal City, Arlington, VA. The reason for this is that the printed patent copies are filed in groups that correspond to the class and subclass classifications. Every patent that is currently classified or cross-referenced in the subclass since the numbering system started should be in that stack of patents. Other locations have the patents filed in numerical order. A listing, available on microfilm or CD-ROM, allowing one to print a list of numbers of the patents in the subclass can be used to check if all are present. (This is necessary for infringement and validity searches to determine if all patents in a subclass have been evaluated.) This listing is also utilized in some of the PDL facilities to enable one to locate patents in a subclass.

Searching in the PTO Search Room has other advantages. It is free. The only charges levied are for copy production and for the use of the major computer systems (which are not needed for hand searching). It is staffed with personnel who have been trained in locating the search areas and who can provide instruction as to the use of the computer systems that are available in the Search Room. Hand searching does not mean that none of the systems and devices using computer technology is avoided. It means that an actual review of the printed paper patents was made. The printed paper patents are individually looked at and evaluated. The PTO has gen-erated many computer databases that are useful and helpful to the searcher. These databases have also been supplied at no charge to the PDLs for use by the inventors who visit the PDL facility in their area. These are known as the CASSIS CD-ROM series and are available to the general public and may be purchased from the PTO. The information available on these disks is limited timewise but provides assistance to the searcher, whether he is a hand searcher or a computer searcher, as to where to look further to find the complete information.

The problem raised in the previous chapter as to the location of the search area is critical to any search. Assuming that one has located the area, how does one proceed? If searching by hand, one locates the cabinet con-taining the class and subclass being sought. Remove the patents from the shelves or trays in which they are placed, take them to a table, and start to go through them in a methodical fashion. Some searchers start with the oldest patent, while others start with the most recent. Look at the first page. If it is a relatively recent patent, there will be an abstract that will define or outline what the patent covers. Do not overlook the drawings and diagrams in the patents. A drawing may indicate that the title and first appearance of the patent did not really tell what it contained. If the drawing is similar to the idea being searched, then the patent may be pertinent, and will require additional study. Having located a pertinent patent, record its number, and in notes, briefly list why it is relevant. If a patent does not appear to be pertinent, go to the next one, and repeat the process until the entire group of patents has been reviewed. In this manner, one will, upon reaching the end of the pile, have a list of the relevant patents that one has located. If additional subclasses must be searched, be sure to list where the patents were found. Later, one will want to obtain copies of those patents for review.

After completing the search of the files and obtaining copies of the patents, it is then time to evaluate each one in relation to the invention being studied. This evaluation should be made by the inventor. If he believes that the patents located do not show his invention, then the attorney or agent who will handle the case should also review the documents discovered.

## Computer searching

In all areas of technology, chemical technology has the best and most advanced indexing practices. The American Chemical Society (ACS), through its *Chemical Abstracts* publication, realized long ago the need for rapid identification and organization of the chemical industry. It was one of the first to provide comprehensive indices to its publication. These indices provide lists of subjects, reactions, conditions, patents (both domestic and foreign), as well as specialized coding systems for specific compounds, which make searching in their publications very easy. With the computer, it was obvious for the ACS to convert these indices to databases and allow the user to search the databases produced. Portions of these databases are open to the public at no charge, while others are available for a fee. The examiners in the chemical area of the PTO use these databases for much of their searching.

If a computer search is to be made, many areas have to be considered. Is it to be done in the PTO facility in Arlington, VA, at a PDL facility near home, at one's employer's office, or on the Internet? Also, another factor to consider is the fact that most computer searches, except some on the Internet, are structured as costing various amounts of money per hour. The current rate at the PTO facility in Virginia is $40 or $50 per hour, depending on the search system used. The exact charge may be prorated based on the minutes actually used. Some PDL facilities will not charge anything, while others will. The fees charged by the PDL facility are based on their costs, and are not controlled by the PTO. Commercial databases usually require a signed contract, and their fees range from $50 per hour to the hundreds and even thousands when the contractual cost is added to the cost of searching.

Computer searching usually requires submitting to the database search engine a word or group of words that describes the invention to be searched. The selection of these keywords or phrases determines what the computer system will produce. The selection of these words may be simple and easy, or very complex and difficult. This depends entirely on the operator and his/her familiarity with the area of technology being searched. The selection of keywords depends on the selection of the search area, as discussed in the last chapter, and is a subject one must learn in order to be an efficient computer searcher.

Remember that a computer does not think. It searches the database being used for exactly what has been typed into the computer. It will only retrieve what is identical to the input operation. As discussed in the previous chapter, spelling is a major problem. Quoting a PTO publication for its examiners, "Not only do we have to be aware of different spellings, we also have to be aware of British terminology." They quote in this document the different use of the terms "hemoglobin vs. haemoglobin," "airplane vs. aeroplane," and "boot or bonnet or windscreen vs. trunk or hood or windshield." They present the difference in the number of responses received from a database. If the PTO stresses this to its examiners, it could go unconsidered by an untrained computer user. This is not a subject to be ignored because one is only searching U.S. patents. Patent applications are submitted from all over the world, and spellings and word usage are even quite different throughout the U.S.

Probably the most important factor to be considered in a computerized search is the scope of the database. Scope means the time period that the database covers. How recent is its last entry, and when did the first patent on the database issue? Very few databases cover all of the numbered U.S. patents. The major database, the PSIS (Patent Search and Image System) constructed by the PTO for use by its examiners, permits full text searching (to be detailed later in this chapter) from 1971 to the latest patents issued. This database is updated weekly as new patents are issued. Prior to 1971, the database allows the user to show on the computer screen the patents in the subclass that is being searched but does not allow for them to be searched by computer for contents. The images of the patents can appear, and the examiner must go through them patent by patent, just as if he/she were hand searching, although they appear on the computer screen. Will these older patents be added to the database to allow for searching by computer techniques? A program is now under way to establish the entire list of patents from 1790 to the current date on DVD-ROMs within the next several years. The present database program is available to every examiner on his/her office computer and is tied to the giant mainframes in the PTO. This database is available in selected PDL facilities, at costs determined by that facility. The PTO is currently charging non-employees $50 per hour.

A second patent data file, the APS (Automated Patent System), was also created by the PTO and is available in the PTO Search Room and at all PDL facilities. It allows searching on a database that extends from 1971 to the current date. It is a database of text only — no graphics or equations — which limits the search. No chemical formulae are shown and no diagrams of circuits are included. This service also bears a fee at the PTO. This database has been in use for some time and is used by many people. It is being updated on a regular basis and is also available for purchase. The PTO sells it on an annual basis, and the cost is about $7000, with renewals at about $5500 per year. This database in the Public Search Room costs $40 per hour.

"Full-text searching" is a term used to indicate that the total contents of the entire patent document are available to the operator for determining if the words or terms asked of the computer are contained in the text portion of the patent, regardless of the portion. Many search systems search for the words or phrases in the title, or abstract, or claims, and generally are programmed to search in this manner. In full-text searching, it does not matter where the term is found; it will be noted by the search system.

Other computer searching sources available to the public can be found on CD-ROMs produced and sold by various commercial manufacturers, including the PTO. These vary in scope from dates in the 1960s, 1970s, or 1980s, and may be updated on a weekly, biweekly, or monthly basis. Also available are many commercial suppliers of patent information that provide online searching through one's own computer. These may allow a one-time search at no charge but normally require a contract covering the costs.

A problem for the computer searcher is to study the scope of the databases that are being used. If one database covers the area of patents from 1980 to current, for example, while others cover the period from 1989 to current times, the results gained from each search must be considered based on the time frames. So, to have a complete search, one must consider the databases, search those most relevant to one's invention, and evaluate if any or all can serve one's purpose.

CD-ROMs are being used by many foreign patent agencies, such as the European Patent Office, the Japanese Patent Office, and the WIPO (a United Nations Organization), but these are mainly being used to transmit patent copies, as the CD-ROM disk can hold many copies of patents and allow the holder of the CD-ROM to print copies of the original patent. It is more economical to use computer disks than paper copies. Each CD-ROM can hold 330,000 typewritten pages on one disk and allow the holder to utilize specialized software to search those disks. The large capacity of CD-ROMs has enabled producers to add the graphics, formulae, and drawings that accompany the patent, and to allow for complete distribution of the contents. For example, a full year's copy of U.S. patents will require about 100 separate disks; this means that many disks must be questioned and studied.

The PTO has announced that beginning in 2000 it will be supplying data to the PDL system and the data will be available for sale to the public in the DVD format, which will mean that the public will be able to get more information on a disk but will require the user to obtain new reading and programming equipment. The PTO will also start supplying U.S. patents in weekly DVD format with the first issue of 2000.

Services are becoming more and more available on the Internet for free searching. International Business Machines has placed a database on the Internet covering the period of 1971 to current issue, at no charge. This Internet source also contains information on European and other foreign sources. The PTO has now placed a database on the Internet that will allow for free searching. This database is similar to the APS system, but each has differing scopes and uses different search engines, which require different methods of querying the search engines. Foreign patent offices have Internet addresses and provide a variety of services.

The commercial databases that currently exist on the Internet will provide rapid responses to the user. But consider their scope. Commercial databases such as Derwent, Dialog, MicroCopy, Corporate Intelligence, and others are broader in scope and may be more specific to a technology area than the free sources. The legal databases (e.g., Lexis-Nexis, etc.) have added patents to their field of search and provide another available data search area. The quantity of databases on the Internet is numerous.

Each of these computerized databases will probably have different searching techniques and methods that should be evaluated before one starts a search. It is advisable to read the HELP section of the databases; most will have advice on how to organize a request. The use of Boolean Search Techniques is not universally adopted by all data providers. Read and copy the HELP sections in order to be sure of how to direct the search engine to find what is desired. For example, using one search system on a patent might not supply a specific patent as a reference, while another search system will. As an example, a patent that employs the principle of scanning fingerprints using a hologram may not be found if the patent uses the term "finger print" as two words and the search system does not combine them to produce the patent in question, although the operator requested "the words together."

Some database producers use the patents as they are received from the issuing agencies, while others add to the material their own abstract and indexing information. Some take the foreign language documents and add a title in English, and others add distinguishing codes to identify patent types or areas and allow these codes to be used to find other similar patents. By creating their own titles and abstracts, these databases allow for better use of keywords and technical content than is given in the abstract supplied in the patent.

A problem that may confront the searcher is the determination of the class and subclass where a patent is currently located. Many of the databases do not adjust their information as to the present classification of the patents but merely include the information as printed on the face of the issued patent. In order to determine the current location of the specific patent, one must, to be sure, go to the CASSIS system and question that database for the current location. See Chapter 8 for additional details.

The PTO places in the examiner's hands the use of many databases for their searching efforts. The PSIS system, Derwent, STN (the total Chemical Abstract database), Dialog, Orbit/Questel, and others are available to the examiner. Which ones the examiner chooses to use is his determination, based on experience and the time he has to devote to the searching process. Some examiners feel that their area of expertise is covered by a specific database and will use it, while others feel that the idea of the invention predates the computerized database and a hand search is the only way to determine if the art presented is novel. Since an attorney or agent handling the application does not know how the examiner searched, he must be prepared for his findings.

Computer searching is at its best in several areas. One is where the invention contains multiple components, such as a cosmetic. By questioning the database for each combination and then narrowing the area, a searcher can quickly find what has occurred in this area and the pertinent patents that have used these components before. Another area is where the technology or application is very new, and unknown before a fixed date (e.g., biotechnology). Then, a computer search using a database having a scope prior to that date should present the total picture.

There are advantages to the use of computer searching. The ability to use computer software to scan a multitude of patents to find the ones that fit the questions asked can greatly shorten the time of a search. As databases are enlarged to contain all patents and new searching machines are developed, the advantages of computer searching will become greater. Until that time comes, there is still a need for hand searching of the patent files. All inventions are not new and novel. Although some may seem very new, prior art can still be found in the patent files of years gone by, and not accessible by a simple computer search. It must be remembered that the term "prior art" covers *all* information, not just patents on the database.

A listing of various Internet sources is presented in Chapter 11, but it is not meant to be a complete listing of all patent sources on the Internet.

## chapter ten

# Patents as legal documents

The issuance of a patent marks the beginning of the inventor's exclusive legal right to the invention and to prohibit anyone from making, using, or selling the invention. Also, upon issuance, the patent becomes statutory prior art as of the filing date of the patent.

The issuance of the patent also gives the inventor the right to mark any product, within the scope of the patent, that the patent holder makes and markets with its patent number. This type of notice is presumptive notice to everyone, whether it is seen or not, that the product is protected by a U.S. patent and is a legal warning to potential infringers.

A patent can be used as an asset. It can be used to stop the competition from using the patented technology. The patent can be licensed to others to make, use, and sell, for a price that is negotiable, or it can be used as a bargaining chip to acquire other technology in a cross-licensing arrangement.

If the patent is going to form the basis for a new product, the patent owner will generally seek an opinion from counsel as to the "right to use" the technology. This opinion will be based on a right to use search (or infringement search) of the patent files to determine if any other patent would restrict its use or the patent owner's freedom to act. If the patent owner is going to market the invention, it is vital to have sound opinions from experienced patent counsel concerning the validity, enforceability, and infringement of the patented invention.

Under U.S. law, infringement of a patent occurs when a person or entity, without authority of the patentee, makes, uses, or sells the patented invention within the United States. Patents are territorial in nature, and it is not generally considered as infringement of a U.S. patent to practice the technology in another country. There is a notable exception. Any product made by a process covered by a U.S. patent, imported into, sold in, or used in the U.S., without authority of the patentee, infringes that patent.

Infringement is determined by attempting to read the claims of the patent on the allegedly infringing device or process. If the language of the patent claim literally reads upon the accused device or process, then the device or process may be said to infringe the claim. Many court decisions have been made concerning the reading of the claim on the device or process, and specific rules have been developed to determine infringement.

Only the claims are enforceable; the specification of the patent can be used to explain and clarify the claims, but cannot enlarge or change them. The claims are very important and should be considered very carefully by the inventor and the individual prosecuting the application.

If the inventor claims less than what is disclosed in the specifications, the patent will only protect the scope of what is in the claims; the remainder will be dedicated to the public. Another rule of claiming, which applies mainly to mechanical claims, states that a claim must relate an operative structure, telling how the various parts cooperate. If this is not shown, the claim may be invalid.

Claims may be either independent or dependent. **Independent claims** are claims that recite or state the critical limitations of the invention in their broadest terms and do not have to be associated with any other claim. The first claim in a patent is an independent claim. The others may be independent or dependent. **Dependent claims** are claims that incorporate, by reference (always to an independent claim), the features of one or more of a preceding claim, with the addition of some further limitation of a specific compound or condition within the scope of the main claim. In other words, the independent claim is a broad claim, while the dependent claim narrows the independent claim. Other dependent claims may also depend on an earlier dependent claim or claims.

A claim can be broken down into three parts:

1. The preamble or introductory phrase
2. The transitional word or phrase
3. The body of the claim

There is no set statutory form for claims. The present PTO practice is that each claim must be the object of a sentence starting with "I (or We) claim," "The invention claimed is," or its equivalent. Each claim begins with a capital letter and ends with a period. Periods may not be used elsewhere in the claims, except for abbreviations. A claim may be constructed with the various elements subdivided in paragraph form.

One can use the patent shown in Figures 10.1 to 10.6 to evaluate the scope of the claims. On the last page of the patent, a set of Claims 1 to 11 is found; some of these claims are discussed here.

Claim 1 (Figure 10.6) states:

1. A rotary film calciner for heating materials by application of indirect heat comprising:
   a horizontally extending rotatable drum having an upstream end and a downstream end;
   a center tube concentrically positioned within the rotatable drum forming an annular combustion chamber

US005906483A

**United States Patent** [19]

**Zhou**

[11]   Patent Number:      **5,906,483**

[45]   Date of Patent:      **May 25, 1999**

[54] **ROTARY FILM CALCINER**

[75] Inventor: **Jun Pei Zhou**, East Amherst, N.Y.

[73] Assignee: **Harper International Corp.**, Lancaster, N.Y.

[21] Appl. No.: **09/071,396**

[22] Filed:      **May 1, 1998**

[51] Int. Cl.⁶ ................................................. **F27B 7/00**

[52] U.S. Cl. ...................... **432/112**; 432/103; 432/113; 432/114; 159/11.1; 159/11.2; 159/16.2

[58] **Field of Search** ...................................... 432/103, 112, 432/113, 114, 118, 98, 102; 159/9.1, 9.2, 10, 11.1, 11.2, 11.3, 12, 16.2, 33, 49; 34/134, 135, 136, 137, 138

[56]                **References Cited**

U.S. PATENT DOCUMENTS

1,944,452   1/1934   Ochs ........................................ 159/9.2
3,343,587   9/1967   Triplett et al. ............................ 159/12

Primary Examiner—Teresa Walberg
Assistant Examiner—Jiping Lu
Attorney, Agent, or Firm—Arthur S. Cookfair; James J. Ralabate

[57]                **ABSTRACT**

An apparatus for drying/calcining comprises a horizontally extending rotatable drum and a center tube, concentrically positioned within the rotatable drum to provide an annular combustion chamber between the outer wall of the center tube and the inner wall of the rotatable drum. One or more gas jet burners are positioned to inject gas fuel and combustion air into the annular combustion chamber in a tangential direction to create a vortex flow of combustion gases through the annular combustion chamber. The flue gas is discharged through the center tube. In operation, a material to be dried and/or calcined is applied to the external wall of the rotatable drum as it rotates toward a scraper or doctor blade where the dried or calcined product is removed by scraping.

**11 Claims, 3 Drawing Sheets**

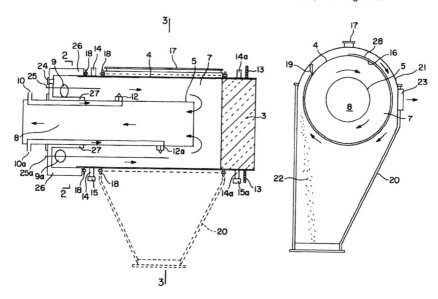

*Figure 10.1*   Title page of U.S. Patent 5,906,483.

between an outer wall of the center tube and an inner wall of the rotatable drum and forming a combustion gas exhaust chamber within the center tube;

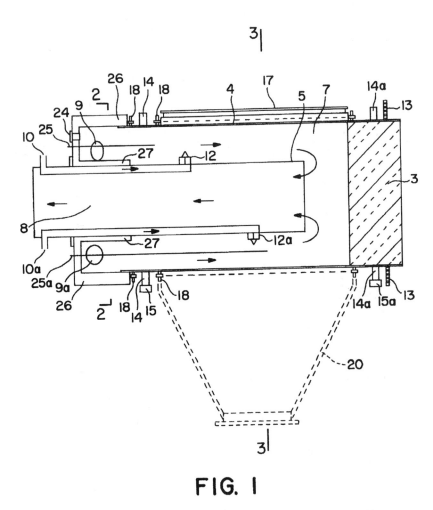

# FIG. I

*Figure 10.2*   First page of drawings for U.S. Patent 5,906,483.

said annular combustion chamber and said combustion gas exhaust chamber each having an upstream end and a downstream end;

said center tube being open at each end to allow passage of combustion exhaust gases therethrough and having

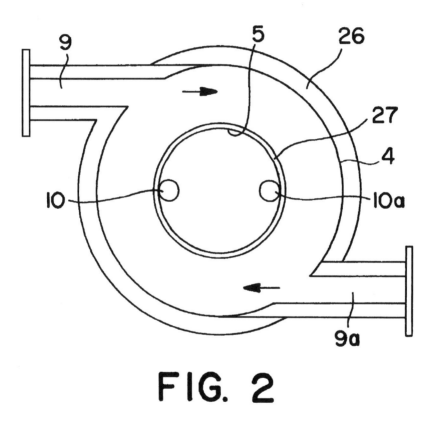

# FIG. 2

*Figure 10.3* Second page of drawings for U.S. Patent 5,906,483.

the upstream end spaced apart from the downstream end
of said annular combustion chamber;

    primary burner means for injecting combustion gases
into the annular combustion chamber and directed to cause
the combustion gases to travel from the upstream end thereof in a spiral path through the combustion chamber and exit
through the gas exhaust chamber;

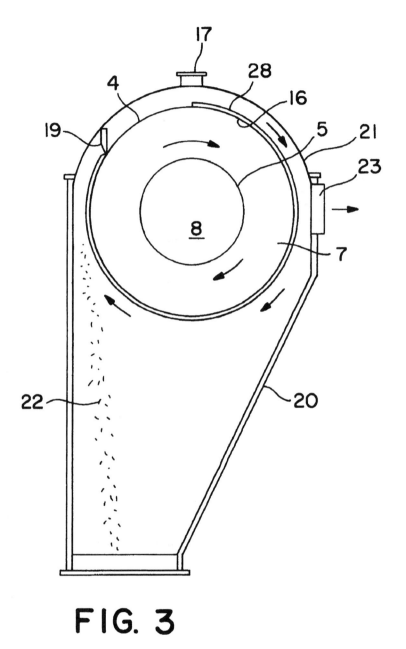

# FIG. 3

*Figure 10.4*   Third page of drawings for U.S. Patent 5,906,483.

5,906,483

**1**

### ROTARY FILM CALCINER

#### BACKGROUND OF THE INVENTION

1. Field of the Invention

This invention relates to a gas-fired rotary industrial process furnace, adapted for the treatment by drying and/or calcining of materials and collection of the treated materials.

2. Prior Art

A variety of industrial process furnaces are known and have been designed with features that provide for the various considerations found in the drying and/or calcining of materials. Such considerations include, for example, the temperature required; whether the furnace should be of the direct-heat type requiring direct contact between the hot gases and the material being treated, or the indirect-heat type wherein heat is transferred by conduction through a wall to the material being treated; whether the operation should be continuous or batch; the particular materials to be treated, the economy of the construction and the operation of the furnace, etc.

U.S. Pat. No. 3,228,454 discloses a drum dryer of the indirect-heat type wherein the inside of a rotating drum is heated with steam while a slurry of the product to be heated is applied to the outside of the drum. As the drum rotates, the product is dried and then removed by means of a doctor blade. The apparatus is used for relatively low temperature operations such as the drying of cereals.

U.S. Pat. No. 5,299,512 discloses a burner designed for a rotary kiln and utilizing a particulate solid fuel, such as coal dust, with a liquid or gaseous fuel. The apparatus comprises concentrically positioned pipes to provide annular channels for the introduction of air, coal dust, and a liquid or gaseous fuel.

U.S. Pat. No. 5,123,361 discloses an annular vortex combustor having an outer vertical exhaust chamber. Fuel, such as powdered coal or coal water fuel, and air are injected tangentially near the bottom of the combustion chamber. Additional air is injected at selected points along the length of the combustion chamber.

U.S. Pat. No. 4,144,019 discloses a burner of the double vortex type wherein combustion gases and particulate travel in a spiral path through an outer cylindrical combustion chamber, then reverse direction to flow through an inner cylindrical combustion chamber and exit therefrom.

U.S. Pat. No. 3,799,252 discloses a roller drier wherein material to be dried is fed to the roller by means of a cylindrical feeding device and the dried material removed by means of a doctor blade.

#### SUMMARY OF THE INVENTION

The present invention provides a vortex combustion rotary film calciner well suited for drying and/or calcining materials by application of indirect heat, and characterized by a combination of important advantages, including high heat transfer coefficient, a wide operable temperature range, excellent uniformity of temperature, high evaporation effectiveness, and energy saving through high combustion and thermal efficiency. The rotary film calciner of the present invention comprises a horizontally extending rotatable drum having a closed end and a center tube concentrically positioned herein to provide an annular combustion chamber between the outer wall of the center tube and the inner wall of the rotatable drum and a combustion gas exhaust chamber within the center tube; at least one gas jet burner positioned to inject gas fuel and combustion air into the annular

**2**

combustion chamber in a tangential direction to create a vortex flow of combustion gas along the length of the combustion chamber with a reversal of direction at the closed end of the combustion chamber to exit through the exhaust chamber. A feed mechanism is provided to apply a material to be dried or calcined to the external wall of the rotatable drum, as it rotates and a scraper or doctor blade to remove the dried or calcined product.

#### BRIEF DESCRIPTION OF THE DRAWINGS

FIG. 1 is a sectional view of a rotary film drier/calciner of the present invention.

FIG. 2 is an end cross-sectional view taken along line 2—2 of FIG. 1.

FIG. 3 is an end cross-sectional view taken along line 3—3 of FIG. 1.

#### DETAILED DESCRIPTION OF THE INVENTION

The calciner of the present invention, as illustrated in FIGS. 1–3, includes a rotary drum 4 having a center tube 5 concentrically positioned within drum 4 to form an annular combustion chamber 7 for passage of combustion gas and an combustion exhaust chamber 8, for the exiting of flue gas. Combustion gas and air are injected into the combustion chamber through burners 9 and 9a. Secondary air and/or gas for staged combustion may be supplied if desired through secondary gas pipes 10 and 10a and injected into the stream of combustion gas through secondary gas injection nozzles 12 and 12a. Burners 9 and 9a are positioned to inject a flow of combustion gas and air in a direction tangential to the combustion chamber to create a spiral flow. Similarly, the secondary gas nozzles 12 and 12a are directed in a manner that will contribute to the spiral flow. The flow of gases follows a generally spiral path from the burners 9 and 9a at the upstream end to the downstream end where, at insulated combustion chamber end 3 the gases reverse direction and exit through exhaust chamber 8 within center tube 5. The combustion gases in the annular combustion chamber 7 are contained by end cap 26 and seal 18 and flange 27 which provides support for center tube 5.

In operation, the drum 4 moves in response to movement of sprocket and chain 13, (driven by drive gear and motor means, not shown) with a rotary motion, supported by metal tires 14 and 14a on trunions 15 and 15a. The material to be treated is applied to the outside surface of the drum by applicator means 17 to form a film 28 on the surface of drum 4. As the drum rotates, the film of material is indirectly heated by the burning gas through the wall 16. Typically, in a drying/calcining treatment, the material is applied, for example, by applicator means 17 through a series of nozzles (not shown) positioned lengthwise above the upper surface of the drum. The material to be treated is generally applied as a wet slurry or paste to form the film 28 on the drum surface and, as the heated drum rotates, the material is first dried and then calcined. The thermally treated material is removed by scraper 19. As the treated material is removed, the stream of removed material 22, typically in the form of flakes or powder, falls through hopper 20, to a collection container (not shown).

In a preferred embodiment, as depicted in FIG. 3, the rotary drum is enclosed to contain and control the volatiles as well as any dust or particulate matter from the drying/calcining treatment. Thus, a shroud or cover 21 is attached to the upper portion of hopper 20 to contain steam, dust, or volatiles from the drying/calcining and/or scraping opera-

*Figure 10.5*  First page of specifications for U.S. Patent 5,906,483.

5,906,483

**3**

tions. Such emissions may be discharged in a controlled manner through off-gas discharge port **23**.

The enclosure of the drum **4** by cover **21** and hopper **20** during the drying/calcining operation provides an additional advantage. Selected gases may be injected within the enclosure to provide an atmosphere compatible with the drying/calcining of specific materials to avoid unwanted chemical reactions.

Various fuels may be employed including, for example, dry finely pulverized coal, coal water fuel, oil, or preferably gas, such as propane or natural gas. The preferred fuel is natural gas. The fuel and air are premixed and ignited and tangentially injected into the combustion chamber. The hot combustion gases, typically at a temperature of about 1000° C., then travel in a spiral path through combustion chamber **7**, reversing direction at insulated combustion chamber end **3**, to exit through exhaust chamber **8** within center tube **5**, providing a uniform distribution of heat along wall **16**. The flame condition inside the combustion chamber may be viewed through view port **24**. The temperature in the combustion chamber may be monitored by means of thermocouples **25** and **25a** and controlled by appropriate adjustments of gas and air being injected.

Although the invention has been described with reference to certain preferred embodiments, it will be appreciated by those skilled in the art that modifications and variations may be made without departing from the spirit and scope of the invention as defined by the appending claims.

What is claimed is:

1. A rotary film calciner for treating materials by application of indirect heat comprising:

a horizontally extending rotatable drum having an upstream end and a downstream end;

a center tube concentrically positioned within the rotatable drum forming an annular combustion chamber between an outer wall of the center tube and an inner wall of the rotatable drum and forming a combustion gas exhaust chamber within the center tube;

said annular combustion chamber and said combustion gas exhaust chamber each having an upstream end and a down stream end;

said center tube being open at each end to allow passage of combustion exhaust gases therethrough and having the upstream end spaced apart from the downstream end of said annular combustion chamber;

**4**

primary burner means for injecting combustion gases into the annular combustion chamber and directed to cause the combustion gases to travel from the upstream end thereof in a spiral path through the combustion chamber and exit through the gas exhaust chamber;

means for rotating said horizontally extending rotatable drum;

applicator means for applying a material to be treated to outer wall of the rotatable drum to form a film thereon;

product removal means for removing treated material from the outer wall of the rotatable drum.

2. A rotary film calciner according to claim 1 wherein the center tube is stationary.

3. A rotary film calciner according to claim 1 wherein center tube is rotatable.

4. A rotary calciner according to claim 1 where in said applicator means for applying a material to be treated to the outer wall of the rotatable drum comprises a series of nozzles positioned above said rotatable drum and extending in a line parallel to horizontal longitudinal axis thereof.

5. A rotary film calciner according to claim 2 wherein said primary burner means comprises two burners tangentially positioned approximately 180 degrees apart within said annular combustion chamber near the upstream end thereof.

6. A rotary film calciner according to claim 5 wherein said combustion gases are formed from natural gas and air.

7. A rotary film calciner according to claim 6, further including secondary gas inlet means for the injection of gases within said annular combustion chamber at a location upstream of said primary burner means.

8. A rotary film calciner according to claim 7 wherein said secondary gas inlet means comprises two gas inlets positioned approximately 180 degrees apart and at different upstream distances from said primary burner means.

9. A rotary film calciner according to claim 5 wherein said product removal means comprises a scraper blade positioned to remove treated product from the outer wall of the rotary drum.

10. A rotary film calciner according to claim 9 wherein said scraper blade is adjustable to compensate for changes in diameter of said rotary drum due to expansion from heat.

11. A rotary film calciner according to claim 9 further including an outer cover and hopper to contain gaseous products and to direct the flow of treated material removed by said product removal means.

\* \* \* \* \*

*Figure 10.6*    Last page of specifications and allowed claims for U.S. Patent 5,906,483.

means for rotating said horizontally extending rotatable drum;

applicator means for applying a material to be treated to the outer wall of the rotatable drum to form a film thereon;

product removal means for removing treated material from the outer wall of the rotatable drum.

The first claim must always be an Independent Claim, and it should be the one that has the widest scope.

In this claim, the **preamble** is "A rotary film calciner for treating materials by application of indirect heat." This preamble states that the invention is a manufactured article, using indirect heat to treat materials.

The **transitional word** is "comprising." The transitional phrase is a "coded" way of stating that the specific or stated components in the claims may have some additional components added to the process. Different transitional phrases have different meanings and are discussed below. In this case "comprising," as used in a claim, denotes that the invention includes the components especially mentioned in the claim, but does *not* exclude others that are not mentioned in the claim.

If the transitional phrase, "consisting of" had been used, the meaning would change to denote that the components mentioned in the claim are the *only* components that can be used. This strictness can be somewhat modified by the inclusion of a modifier such as "essentially" to read "consisting essentially of." This modification has been interpreted to mean that the addition of a minor amount of an additional component that would not affect the result or properties of the result would be allowable.

**The body** of the claim, coming after the transitional phrase states, usually in a series of phrases, the components or elements or steps that form the invention. In this example, the body is quite lengthy, composed of eight phrases. Several of these phrases will be explained to illustrate the construction of a claim body. Remember, the rules require that the claim (together with the introduction, "What is claimed is:") be a single sentence, so the sum of the phrases must define the invention. If one looks at the first phrase, "a horizontally extending rotatable drum having an upstream end and a downstream end," this phrase starts to establish the structure of the unit, as well as describing the path of gases through the unit. The next phrase, "a center tube concentrically positioned within the rotatable drum forming an annular combustion chamber between an outer wall of the center tube and an inner wall of the rotatable drum and forming a combustion gas exhaust chamber within the center tube," further establishes the structure of the unit and its position in the unit, as well as providing names for elements in the unit. (These elements will be modified in later sections of this claim, as well as in some of the other claims.)

The sixth phrase, "means for rotating said horizontally extending rotatable drum," uses the word "means," which is a way of stating that a method of driving or turning the drum is accomplished by some method, which is not specific to this unit and would cover any attempt to circumvent this patent by stating a specific method of applying the energy to cause rotation. The use of "means" will be found in many patents to cover methods of machinery or techniques that are conventional to the operation involved. In a similar fashion, the use of "means" in the last phrase, "product removal means for removing treated material from the outer wall of the rotatable drum," indicates that some method may be necessary, such as a scraper or knife or other device, to remove the material from the wall and allow it to be collected.

Claim 2 states:

> "A rotary film calciner according to claim 1 where in the center tube is stationary."

This is a Dependent Claim and places limitations on claim 1 by restricting the "center tube" as described in claim 1 as stationary, or not rotatable. This claim incorporates all the language of claim 1, as well as the restriction placed on the center tube by claim 2.

Claim 5 states:

> "A rotary film calciner according to claim 2 wherein said primary burner means comprises two burners tangentially positioned approximately 180 degrees apart within said annular combustion chamber near the upstream end thereof."

This is a Dependent Claim that encompasses all of claim 2, which in turn, is dependent on claim 1, and incorporates additional restrictions to the unit. Note the use of "means" to broadly cover the type of the primary burner, and the use of the term "comprises" as described above to describe the number of burners and their placement in the chamber.

Claim 9 states:

> "A rotary film calciner according to claim 5 wherein said product removal means comprises a scraper blade positioned to remove treated product from the outer wall of the rotary drum."

This is a Dependent Claim, wherein additional limitations are added to claim 5, which includes the limitations added in claim 5, as well as the limitations added in claim 2 to the contents of claim 1. The limitation of claim 9 is a scraper blade to remove the product from the drum.

These claims show how the initial claim was more specifically narrowed, from the broader description of the invention through the incorporation of the dependent claims accepted in this case.

There are different types of claims, and each serves its own purpose; but in evaluating any claim, read it carefully, observe the punctuation, observe the paragraphing and numbering of steps or segments, and evaluate the transitional phrases very closely.

A patent has been awarded, and the inventor understands what he has invented and how it was claimed, and believes that he knows his rights. Will the courts agree with him and vigorously enforce his rights against infringers? Since 1982, the courts have permitted enforcement of patent rights in about 70% of the cases. Prior to 1982, the courts were holding about 60 to 70% of the patents invalid. This turnaround occurred through the creation of the U.S. Court of Appeals for the Federal Circuit (CAFC), which first convened in October 1982. The legislation that founded the CAFC also established that all patent cases shall come to that court. The strength and value of patents have increased dramatically because the CAFC understands the technical side of the patents. This court has instilled throughout the entire federal court system a uniform application of the patent laws and has strengthened the enforceability of the patent grant. Patents have been restored as a valuable asset to industry and are carefully evaluated by patent owners and prospective infringers alike. Licensing of patents has increased, and the advantages of patent protection are now a major consideration of industry.

The cost of defending a patent, or the cost of stopping an infringer from using a patent, can be very high. This fact is sometimes used to "scare" independent inventors or small companies away from legal action to enforce their patent grant. Neither the size of the corporation nor the number of attorneys makes the decision; the law is still applied fairly and evenly. Justice will prevail!

# chapter eleven

# Patent information
# from the Internet

The expanding use of the Internet as a vehicle for information retrieval has included the establishment of a large number of Web Sites that pertain to the field of patents. Many of these offer pertinent and relevant information. Others are merely advertising for products or services connected to the patent field. As mentioned in a previous chapter, the search engines provided and used by the various Internet Service Providers (e.g., AOL®, Prodigy®, etc.) will search the Web Sites in different ways. If one enters the phrase "patent searching," the results of a specific search may be completely different.

Before assuming any Web Site to be the best basis for a search, consider the background of the provider. Does that provider desire to receive recompense? Some Web Sites, such as the PTO site and the Canadian PO site, provide information as a public service. Corporations such as IBM are providing the site and generating profits for the supplier of patents that they promote. No one must buy patents from that provider, and the Web Site will continue.

Many sites will have links to other sources of information. Check these out, as they may prove helpful. The following list of Web Sites and their addresses is not complete but is representative of what can be found and used. The list was accurate at the time it was prepared but may not be the most current address as you are reading. Remember, in this list, the address is given for the URL (Uniform Research Locator). Not all Internet addresses are on the WWW (World Wide Web), so insert the address following the http://indication.

The list has been prepared with the reader in mind and is grouped into the following categories:

## United States Patents
### General Information

**U.S. Patent and Trademark Office**    **www.uspto.gov**
This should be the starting point for all persons who want to obtain data or information on U.S. patents and trademarks. This database will allow for patent and trademark searching, as well as provide background data on patents and trademarks; what constitutes a patent application; how to obtain the forms needed for filing and prosecution; and other useful information.

### Other U.S. Government Branches and Agencies

**U.S. Court of Appeals for the Federal Circuit**
**(CAFC)**                  **www.fedcir.gov**

**U.S. Copyright Office**         **lcweb.loc.gov/copyright**

**Code of Federal Regulations**     **www.access.gpo.gov/cfr/index**

**U.S. Code**                     **law.house.gov/usc**

### Patent Searching
#### U.S. Patents:

**U.S. Patent and Trademark Office**    **www.uspto.gov**

**IBM Patent Server**            **www.patents.ibm.com**

*Chemical Abstracts*           **www.cas.org**
This will enable one to determine which databases are available for use in Chemical Searching, including Chemical Plus and others available from the American Chemical Society.

**Corporate Intelligence**        **www.corporateintelligence.com**
This database will also allow for Trademark Searching.

**Derwent**                   **www.derwent.co.uk**
Databases for worldwide searching are available.

**Dialog Corp.**             **www.dialog.com/info/products**

**FIZ Karlsruhe**          **www.Fiz-Karlsruhe.DE**
This German corporation provides access to many databases in Europe and worldwide. They handle the ACS services in Europe.

**Lexis-Nexis**             **www.lexis-nexis.com**
Primarily a general legal database but covers patent and trademark litigation and decisions.

**Micro Patent**            **www.micropat.com**

**Questel-Orbit**           **www.questel-orbit.com**

**RAPRA Abstracts**       **abstracts.rapra.net**
This database is prepared by the Rubber & Plastics Research Association, and is quite thorough and specific to this field. It is also available on Dialog.

### Patent Organizations and Publications

**American Intellectual Property**
**Law Association**           **www.aipla.org**

**Australasian Legal Information Institute**    austlic.law.uts.edu.au

**Electronic Freedom Foundation**     www.eff.org/pub/intellectual_property

| | |
|---|---|
| Intellectual Property Owners Assoc. | www.ipo.org |
| National Association of Patent Practitioners | www.napp.org |
| *Journal of the Patent and Trademark Society* | www.jptos.org |
| Patent Office Professional Society | www.popa.org |
| *Patent World Magazine* | www.ipworldonline.com |
| *IP Magazine* | www.ipmag.com |
| Intellectual Property Today *Law Works Magazine* | www.lawworks-iptoday.com |
| Inventors Digest Magazine | www.inventorsdigest.com |
| Patent Café Magazine | www.patentcafe.com |
| Patent Law Sources: Academic | |
| Access — Intellectual Property Law | www.access.iplaw.com |
| Indiana University Law School WWW Law | www.law.indiana.edu/law/lawindex |
| Legal Information Institute Cornell University | www.lw.cornell.edu/topics/patent |
| Federal Circuit Cases from Emory University | www.law.emory.edu/fedcircuit |
| Franklin Pierce Law Center | www.fplc.edu./tfield/order |
| Georgetown University Law Center | www.ll.georgetown.edu |
| Villanova University School of Law | www.law.vill.edu |
| Washburn University School of Law | lawlib.wuacc.edu/washlaw/lawjournal |
| Foreign Patent Offices: | |
| Australia | www.aipo.gov.au |
| Austria | www.ping.at/patent/index |
| Belgium | www.european-patent-office.org/country/Belgium |
| Brazil | www.bdt.org.br/bdt/inpi |
| Canada | www.cipo.gc.ca |
| China | www.cop.cn.net/ |
| Croatia | gogan.srce.hr/patent/index |
| Denmark | www.dkpto.dk |
| European Patent Office | www.epo.co.at.ipo |
| Finland | www.prh.fi |
| France | www.inpi.fr |
| Germany | www.deutsches-patentamt.de |
| Greece | www.epo.co.at/epo/patlib/country/greece/index |

| | |
|---|---|
| Hong Kong | www.houston.com.hk/hkgipd |
| Hungary | www.hpo.hu |
| Italy | www.epo.co.at/epo/patlib/country/um_noit |
| Japan | www.jpo-miti.go.jp |
| Korea | www.ik.co.kr/kopatent |
| Lithuania | www.is.lt/opb/engl |
| Luxembourg | www.etat.lu/ec |
| Malaysia | kpdnhg.gov.my/ip |
| Monaco | www.epo.co.at/epo/patlib/country/monaco/index |
| New Zealand | www.gov.nz/ps/min/com/patent |
| Peru | ekeko.rep.net.pe/indecopi/indecopi |
| Poland | saturn.ci.uw.edu.pl/up |
| Portugal | www.inpi.pt |
| Romania | www.osim.ro |
| Slovenia | www.sipo.mzt.sl |
| Spain | www.eunet.es/interstand/patantes/index |
| Sweden | www.prav.se/prving/front |
| Switzerland | www.ige.ch |
| United Kingdom | www.netwales.co.uk/ptoffice/index |
| WIPO | www.wipo.int |

**General Commercial Sources**

| | |
|---|---|
| American Patent and Trademark Law Center | www.patentpending.com |
| Fenwick and West | www.fenwick.com |
| Kremblos, Foster, Millard & Pollick | www.pattorneys.com |
| Kuester Law | www.kuesterlaw.com |
| Ladas and Parry | www.ladas.com |
| Oppedahl and Larson | www.patents.com |
| Patent Information Users Group | www.piug.org |
| Thompson and Thompson | www.thompsonconsulting.com |
| Townsend | www.townsend.com |
| Wells, St. John, Roberts, Gregory and Matkin | www.wstgm.co |

**Inventor Organizations**

| | |
|---|---|
| Listing of State Organizations | www.inventorsdigest.com |

# chapter twelve

# Importance of record-keeping

The recording of data is an essential part of experimentation. In laboratory courses, science and engineering students are generally required to keep records — usually in a laboratory notebook. After graduation, in research laboratories or other scientific or technical work environments, this practice is continued. However, in the work environment, in addition to its obvious importance for scientific purposes, good record-keeping takes on added importance for legal purposes. In industrial research laboratories, researchers are generally required to keep a daily record of their experiments in a laboratory notebook. The notebook then provides an important legal as well as technical record of the work that was done and the experiments performed. Properly kept, the laboratory notebook may later serve as legal evidence to establish rights to discoveries made in the laboratory.

The U.S. patent system is based on the premise that the patent is awarded to the first inventor. The person who can *prove legally* that he/she completed the invention first is the inventor and entitled to the patent. This is why one's notebook is so important to the patent system. It is a method of proving the dates and facts!

Employers generally provide a notebook and a set of record-keeping rules for their employees. The notebooks are bound, and the pages numbered. Some notebooks may contain some printed material on each page, such as a place for the author's and witnesses' signature and space for dates. The binding of the pages is important to show the permanency of the record, as well as the continuity of the recording. The notebook will become the permanent record of the experimentation and will become part of the history of the project. It will be written as the project develops, so it must be a clear and logical record. The recordings may be easily understandable when they are written; but remember, if a patent problem arises, it will not arise now, *but years into the future.* The entries in the notebook should record the information clearly so they can be understood, not only today, but many years from now.

Therefore, as simple logic suggests, when placing entries into the notebook, use ink for all entries. Never make erasures. Write as clearly as possible; do not use acronyms or abbreviations or code words; and as the author of the page, place your signature and the date at the bottom of each page.

These are only a few of the rules or suggestions that will be made to help the reader in the recording of an experiment. With today's instrumentation based on graphs or photos, such documents can be placed in the notebook to eliminate writing or copying the data. Depending on their size, they can be pasted directly to the page or folded and attached to the page. Indicate in the written text that a photo, graph, or chart has been placed on the page, and initial and date the insert (if possible do so on the seam of the insert and page).

Depending on the company's rules, it is good practice at the end of the day to finish off the page by drawing a line from the last word to right side of the page, and then a line connecting that point with the bottom left of the page to indicate that all entries are finished on that particular page. The page should then be signed and dated to signify that the page was completed on the day of the signature.

The recording of critical information and observations may have fulfilled an important step toward establishing the right to a patent. The next step is to have someone who is skilled and knowledgeable in the area witness the written pages. The purpose of the witnessing is to corroborate that the record existed on the date that the witness read and understood what was written. The witnessing of the page should be done as soon as possible after the signing by the author. In some corporations, two witnesses are desired for the corroboration process.

Other employer's rules may mandate that upon the end of a day's recording, the author shall sign at that point, regardless of where it is on the page, and the witnesses will also sign at that point. The next entry will start on the next available line, not starting a new page.

Whichever system is used, be consistent, and use it until the end of the book. If one wishes to change, do it when starting a new book; never change within a book, as this raises the question, was the system changed to "alter" the data, and does it reflect upon the author's veracity?

This same logic applies to errors one might make in writing. Do not erase or obliterate the error. Just run a line through the error and rewrite the correct word or data. It is best to be able to read the error, so it can be clearly seen why the correction was made. The page may not look as neat; but if the page is ever needed in a court of law to prove a legal point, the author will be pleased that the error is visible and the correction is obvious. This may help avoid being asked on the witness stand, "What did you cross out, and why?" A simple reminder: always sign the pages in the same way. Adopt a standardized signature and do not vary it from one time to the next.

Remember, if inventive work and the dates on which it was carried out have to be proven, the notebook may well be the critical evidence of the dates and results. If the problem exists that two or possibly more people claim to have made the same invention, the Patent Office will be forced to determine who is the first inventor. In order to do this, an Interference Proceeding is started; and if you are one of the inventors, you will be involved. You will be questioned *under oath* by an attorney representing the other party or parties, as well as an attorney on your side. This questioning can get rough. If the validity of the patent is called into question after issuance, then your questioning may be conducted in open court. Each of these proceedings can be quite intimidating for the inventor and his witnesses, as well as being quite costly for the patent holder.

Since it is obvious what purpose the notebook will serve in these proceedings, it is best to prepare for them should the Interference Proceedings occur. Although these proceedings are not common, one never knows when they will occur; thus, *all notebook entries* must be ready to withstand critical review at a later date.

What should be recorded in the notebook? The details of the experiments conducted. In addition, the experiment should be identified as belonging to a certain overall program. Also add pertinent facts as to reasons for the experiment and if conducted in an area different from one's usual workplace.

Entries should disclose the observations, measurements, apparatus used, conditions, and results. The degree of completeness should be high, as all information may be needed to allow someone else, if necessary, to duplicate the experiment at a later date. Never use code words for ingredients or reactants; use their correct chemical or technical name.

The entries in a notebook should be in chronological order; so, indicate if this experiment or observation is connected to previous experiments. This is very important, as it shows that a series of experiments was being conducted in an orderly manner, showing diligence, which may be necessary to prove the critical dates of invention, conception, and reduction to practice.

Never place in the notebook comments concerning the results of an experiment, such as "it is a failure" or "it did not work." Never place disparaging comments in a notebook. Never indicate that the work is going to be abandoned or terminated. Simply set forth the data obtained without opinion.

The testimony of the inventor and his records alone are *not adequate* in a court of law to establish inventorship. They must be corroborated by others. This is what the witnessing aims to prove. The more the testimony can be corroborated, the better will be the proof that the information on the written page is true.

The testimony by an independent third-party witness might provide corroboration, but the courts have stated that corroboration requires a "cohesive web of tangible evidence." Tangible evidence refers to evidence in physical form. In contrast, testimony alone is not tangible; neither is memory. The Court of Appeals for the Federal Circuit (CAFC) has adopted a *rule of reason* that says that no specific type or amount of corroboration is required, so long as the evidence submitted forms a "cohesive web of tangible evidence." This rule of reason has been established for several reasons, as no single type of corroboration can be established for all cases and the corroboration must be flexible to allow different factual circumstances to be explained. The PTO will consider any evidence that is submitted and will decide each case based on the facts of that particular case.

The person who witnesses the entries in the lab notebook should not be a co-inventor, but should be a person who has read and understood the experiment. One's lab technician may be a good witness, provided the lab technician is not a co-inventor. Remember, the witness must have read and understood the entries and should be a person who can reasonably be expected to be available for several years after the date of the entry.

The procedure of interference is complex, and only several facets will be explained. The supplying of the earliest date is not the only question that may be raised. Diligence may have to be shown, as well as proper inventorship. Do the records constitute the proof of the entire invention? One's notebooks are now like gold!

There will be many pages of information in one's notebooks concerning the experiments that resulted in the invention. These entries should cover the time period from conception to reduction to practice. There will be examples, in that period, that did not perform as hoped; and there will be others that were total failures. Each is important, as it shows that one was working on the invention but had not found the proper answer as yet. These experiments show diligence on one's own part to solve the problem! This is very important as proof of the earliest date. Diligence may be shown in many ways: by a series of experiments, or statements that "new experiments are planned as soon as reagents arrive," or a plausible reason why one is not working actively on the invention.

Patent interference proceedings occur in about 1% of the issued patents and cannot be determined in advance. All one can do is be prepared in the event that one's invention is part of the 1%. One must train oneself in the proper manner of record-keeping and hope for the best.

Today, almost every laboratory and/or office has a computer terminal, and scientists are using them for many purposes. Some computers are directly tied to equipment and recording devices, and collect and store the information on magnetic surfaces. Why can one not use these records instead of being forced to transcribe them so they can be placed in notebooks? Why can the techniques of these electronic machines not be used to replace the notebooks?

There are many good reasons why this cannot be done today. The main reasons are: permanence, accuracy, and protection against loss and alteration. The reason for keeping accurate notebooks is to store original data in a manner such that the data can be retrieved at a later date and serve as the basis for proof, in a legal manner, of when specific incidents or events occurred. Storing this data on a magnetic disk or tape raises problems. Who wrote the information? Who is the author? Has the disk or tape been modified since its original writing? How has the disk been stored to avoid magnetic fields that could erase or alter the stored information? Could the electronic record have been purposely altered at a later date?

Research records kept on computers can be used as evidence in an interference proceeding, but the question that is always raised is, how does one prove that the record is the true record? How does one prove that the magnetic record has not been modified or altered to suit the present situation? Under the PTO rule of reason, can the inventor prove that the record presented has been protected and guarded by a method that ensures that the set of computerized files is the true record of the experiments?

The procedures and operation of a secured filing system must have been established previously, and there must be sufficient proof that the system is secure and unavailable to tampering or access by those scientists who provide input to the system. Proof must be available that the procedure or technique has been tested to prove that a computer disk or tape has not been modified since its writing. Proof must be available that the entries were written on a specific date and that the disk has not been corrected. Perhaps a larger question to be answered is, how does one witness an entry? These are all legal questions that are being raised throughout industry as more computer-generated data are being stored. Although some corporations are attempting to find a system whereby computer-generated and stored information is protected against erasure, alteration, and addition, and which can assure that the dates are correct, the positive answer has not yet been found.

Although more and more information is being stored in computers and the ease of retrieval of the information is improving almost on a daily basis, the security and permanence of the recording media are major problems in solving the legal questions. With science and technology moving so rapidly, a system may be developed that will answer all of the problems and produce a computer program that will, upon the pressing of a key, produce the notebook in a form that meets all of today's legal objections. But until that day arrives, we must keep our handwritten notebooks and follow all of the rules previously set out.

# chapter thirteen

# Patents around the world

Patents are territorial, enforceable only within the boundaries of the issuing country. If patentable technology is to be commercialized on an international basis, business considerations may dictate the filing of foreign patent applications — a process that is both time-consuming and expensive. Patent prosecution efforts as well as government and legal fees may have to be repeated for each country in which a patent application is filed.

In today's economy, business operates on a global basis, and a product or process can be marketed all over the world. Each product has competition, and one must be ready to cope with the problems of someone trying to use one's process or sell a product that is the same. A U.S. patent will protect the holder from manufacture, use, and/or sale of infringing products within the U.S. This includes infringing products being imported into the U.S. from abroad and infringing products that are manufactured in the U.S. However, a patent has no application where both the manufacture and sale occur outside the U.S. Therefore, the owner of the patent must consider the following questions:

- Are there markets for the products using the invention in any countries, or might one possibly exist in the future?
- Are there potential competitors or potential licensees located in that market?
- Are there significant possibilities of use of the product of the invention in those markets?

If the answer to any of the above is positive, then filing an application in those countries where there appears to be a potential problem or market is a procedure that should be considered.

Although the costs are high, the losses to the patent holder may be even higher, and consultation with a patent counsel skilled in overseas filings should occur. The value of a patent is an asset that must be considered.

From time to time, one reads about some governmental act somewhere in the world that is claimed to lead to a Worldwide Patent. Since there are over 100 countries issuing patents today, one can imagine the efforts that would be necessary to create a single set of rules and laws to produce a document that is valid and enforceable in all of those countries. That is probably an impossibility, at least for the near future.

For more than 100 years, the major patenting nations of the world have sought cooperative approaches that would minimize the problems of seeking patent protection on a global basis. One of the earliest cooperative actions was the Paris Convention of 1883, which provides that patent applicants from any member country can file corresponding applications in other member countries and have the same patent rights as citizens of those countries. Another important provision of the Paris Convention is that, for a priority period of 1 year after a patent application is filed in a Convention country, corresponding applications can be filed in other Convention countries and still enjoy the benefit of the first filing date. Most countries are members of the Paris Convention and the membership includes most of the major industrial nations of the world.

Recent years have seen considerable cooperative effort aimed at the creation of multinational patent systems based on the harmonization of worldwide patent laws. The Patent Cooperation Treaty (PCT), which came into force in 1978, provides a system for filing a single international application and designating member countries to which the application is aimed to be directed. The PCT is administered by the World Intellectual Property Organization (WIPO), a part of the United Nations. Following the filing of the international application, which can be done at the U.S. Patent and Trademark Office, preliminary operations are conducted. A prior art search is made; the designated countries are notified; and certified copies of the documents are prepared. Amendments modifying the claims may be made before the applicant has to decide whether or not to file in the selected countries. Also in this time period, the search report should have been received from the PCT, from the initial application country, and possibly a first office action. This helps in determining how to proceed, especially with regard to foreign filing. Also, new information or a new invention may have arisen that may affect the value of the first invention and change the decision as to filing. If foreign applications are to be filed, the time has arrived for the applications to go into the regular national application phase for each of the selected countries, with the necessary cost of filing fees, translation costs, agents' fees, etc., becoming payable at this time.

The WIPO has issued a list of all countries in the world, which is shown is Figure 13.1, and a two-letter code to identify each country. Not all countries on this list issue patents, but many do, and may qualify to join the PCT program. As of December 1, 1999, the list of member countries had reached 104.

The PCT allows the applicant to file the initial paperwork in English and simplifies the documentation needed for foreign filing. Moreover, it permits the applicant to defer the final decision on foreign filing for a longer period of time and to have the benefit of an official search report prior to making that decision.

WIPO STANDARD COUNTRY CODES

| Country | Code | Country | Code | Country | Code |
|---|---|---|---|---|---|
| Afghanistan | AF | Egypt | EG | Malaysia | MY |
| African IPO | OA | El Salvador | SV | Maldives | MV |
| Albania | AL | European | EP | Mali | ML |
| Algeria | DZ | Equat Guinea | GQ | Malta | MT |
| Andorra | AD | Estonia | EE | Mauritania | MR |
| Angola | AO | Ethiopia | ET | Mauritius | MU |
| Anguilla | AI | Falkland Islands | FK | Mexico | MX |
| Antigua | AG | Fiji | FJ | Moldova | MD |
| Argentina | AR | Finland | FI | Monaco | MC |
| ARIPO | AP | France | FR | Mongolia | MN |
| Armenia | AM | Gabon | GA | Montserrat | MS |
| Aruba | AW | Gambia | GM | Morocco | MA |
| Australia | AU | Georgia | GE | Mozambique | MZ |
| Austria | AT | E. Germany | DD | Mynamar | MM |
| Azerbaijan | AZ | W. Germany | DE | Namibia | NA |
| Bahamas | BS | Ghana | GH | Nauru | NR |
| Bahrain | BH | Gibraltar | GI | Nepal | NP |
| Bangladesh | BD | Greece | GR | Netherlands | NL |
| Barbados | BB | Grenada | GD | Neth Antilles | AN |
| Belarus | BY | Guatemala | GT | New Zealand | NZ |
| Belgium | BE | Guinea | GN | Nicaragua | NI |
| Belize | BZ | Guinea Bissau | GW | Niger | NE |
| Benin | BJ | Guyana | GY | Nigeria | NG |
| Bermuda | BM | Haiti | HT | Norway | NO |
| Bhutan | BT | Honduras | HN | Oman | OM |
| Bolivia | BO | Hong Kong | HK | Pakistan | PK |
| Botswana | BW | Hungary | HU | Panama | PA |
| Brazil | BR | Iceland | IS | Papua New Guinea | PG |
| Brunei | BN | India | IN | Paraguay | PY |
| Bulgaria | BG | Indonesia | ID | Peru | PE |
| Burkana Faso | BF | Iran | IR | Philippines | PH |
| Burma | BU | Iraq | IQ | Poland | PL |
| Burundi | BI | Ireland | IE | Portugal | PT |
| Cambodia | KH | Israel | IL | Qatar | QA |
| Cameroon | CM | Italy | IT | Romania | RO |
| Canada | CA | Jamaica | JM | Russian Federation | RU |
| Cape Verde | CV | Japan | JP | Rwanda | RW |
| Cayman Islands | KY | Jordan | JO | Saint Kitts and Nevis | KN |
| Central Africa | CF | Kazakhstan | KZ | Saint Lucia | LC |
| Chad | TD | Kenya | KE | Samoa | WS |
| Chile | CL | Kiribati | KI | San Marino | SM |
| China | CN | Dem Korea | KP | Sao Tome and Principe | ST |
| Colombia | CO | Rep of Korea | KR | Saudi Arabia | SA |
| Comoros | KM | Kuwait | KW | Senegal | SN |
| Congo | CG | Kyrgyzstan | KG | Seychelles | SC |
| Costa Rica | CR | Laos | LA | Sierra Leone | SI |
| Ivory Coast | CI | Latvia | LV | Singapore | SG |
| Croatia | HR | Lebanon | LB | Slovak Republic | SK |
| Cuba | CU | Lesotho | LS | Slovenia | SL |
| Cyprus | CY | Liberia | LR | Solomon Islands | SB |
| Czech | CS | Libya | LY | Somalia | SO |
| Czech Rep | CZ | Liechtenstein | LI | South Africa | ZA |
| Denmark | DK | Lithuania | LT | Soviet Union | SU |
| Djibouti | DJ | Luxembourg | LU | Spain | ES |
| Dominica | DM | Macao | MO | Sri Lanka | LK |
| Domin Republic | DO | Madagascar | MG | St. Helena | SH |
| Ecuador | EC | Malawi | MW | St. Vincent & Grenadines | VC |

*Figure 13.1* WIPO list of countries and their codes.

WIPO STANDARD COUNTRY CODES

| | | | | | |
|---|---|---|---|---|---|
| Sudan | SD | Trinidad & Tobago | TT | Uzbekistan | UZ |
| Surinam | SR | Tunisia | TN | Vanuatu (New Hebrides) | VU |
| Swaziland | SZ | Turkey | TR | Vatican City State | VA |
| Sweden | SE | Turkmenistan | TM | Venezuela | VE |
| Switzerland | CH | Turks & Caicos Islands | TC | Vietnam | VN |
| Syria | SY | Tuvalu | TV | Virgin Islands | VG |
| Taiwan | TW | Uganda | UG | WIPO | WO |
| Tajikistan | TJ | Ukraine | UA | Yemen | YE |
| Tanzania, United Republic | TZ | United Arab Emirates | AE | Yugoslavia | YU |
| Thailand | TH | United Kingdom | GB | Zaire | ZR |
| Togo | TG | United States of America | US | Zambia | ZM |
| Tonga | TO | Uruguay | UY | Zimbabwe | ZW |

*Figure 13.1*   (continued)

The European Patent Convention (EPC) represents another multinational cooperative effort aimed at simplifying the process of seeking international patent protection. The statement that a patent is only enforceable within the territorial boundaries of the country that issues it is correct; but in this case, it must be interpreted to mean within the borders of the members of the European Patent Convention. In 1977, the member countries established the European Patent Office (EPO) with the power to establish a patent organization that would issue patents that would be honored in all member nations. At that time, 12 countries constituted the EEC, all were members of the Paris Convention, and the EPO was born. Within a few years, it started issuing patents. It is important to note that the individual members of the EPC did not eliminate their own national patent systems, so a patent issued by the EPO could also be issued by the individual member countries. Which patent is more important? The question remains unanswered, as the courts in the member countries have issued rulings that confuse the issue rather than settle the question. Since 1977, additional countries have joined the EPC and more countries have applied for membership.

As time progresses and more countries join the EPC, and the European Free Trade Association (EFTA) becomes more active and larger in scope, it has been planned to issue a single patent, enforceable in all member countries. This patent would have common rights in all member countries. Several problems must be solved prior to the start of this program. Each member country will have to change its patent laws to become uniform, and an agreement must be reached on the number of languages that will be used. It is believed that 11 different languages will be needed; and if so, any application will require translations into all of the different languages. Also under this proposed system, a uniform set of maintenance fees will be established. Proponents of this system claim that the total costs of fees and translation costs will be less than the current costs, but this is yet to be settled.

Every country in the world has its own rules, regulations, and limitations on the patents that it issues. Generally, most countries grant patents for a period beginning on the filing date and usually ending 20 years from that date. Many countries require that the patents be "worked" or the patent will be declared invalid. What constitutes "working" varies from country to country. Some require that the invention be actually manufactured or sold in the country; others require availability of licensing; while still others demand mandatory licensing.

The decision whether or not to attempt to gain foreign patent coverage, as stated above, is based on many factors. The decision to file and where to file is a time-consuming exercise in which one must consider many aspects and also guesses as to the future.

The past decade has seen a strong international movement toward "harmonization" of patent laws. Proponents of harmonization claim that it would simplify and equalize protection and enforcement of patents throughout the world. In 1984, the WIPO called a meeting to discuss the subject. Since then, efforts have been under way to draft a treaty that would fulfill the goal of patent harmonization throughout the world. Progress has been made on this approach, but the end is not in sight. All of the major industrial countries of the world have to sign the treaty, committing each to a series of changes in its own patent laws and rules and placing all inventors on an equal status, regardless of country. Such a program will not occur overnight, as each country is very protective of its own patent laws and rules. This is evidenced here in the U.S. by the attempt to create a program for the publication of all patent applications by the end of 18 months after application date. Opposition arose, lobbyists arrived, and the legislation was "shelved," although the Patent Office, patent law associations, and industrial supporters had all agreed to the proposal.

One aspect of harmonization in the U.S. has occurred as a result of the NAFTA and GATT agreements. The patent law for over a century had established the term of a U.S. patent at 17 years from its issue date. As a result of GATT, the patent term of a U.S. patent changed from the 17-year period to a 20-year period from the date of the application. All patents that were in force on June 8, 1995, or that were issued on an application filed before that date, will automatically have a term that is the greater of 20 years from the filing date, or 17 years from the issue date. The 20-year term period can be extended, no more than 5 years, for delays in the issuance of the patent due to interference, secrecy orders, and/or successful appeals to the Board of Patent Appeals and Interferences or the federal courts.

The GATT-implementing legislation also created a new type of patent application (discussed in Chapter 6) — the Provisional Application.

The U.S. patent system grants its patent to the first inventor to invent, not the first to file the application. When two or more parties claim an identical invention, the PTO commences an interference proceeding to determine the first inventor. In these interference proceedings prior to 1993, a U.S. inventor had a distinct advantage if the other applicant of the invention was a foreigner. Foreign inventors could not establish a date of invention based on activity in a foreign country (other than their date of filing a patent application). U.S. inventors, however, could rely on proof of invention based on their inventive activities before their filing date. As a result of the NAFTA treaty, U.S. laws were amended to extend to Canadian and Mexican applicants the right to rely on proof of inventive activities in their respective countries. With the passage of the GATT-implementing legislation, this right has been extended to applicants from all of the GATT countries that have joined the World Trade Organization and includes the major industrialized countries of the world.

Interference proceedings have always been time-consuming, costly, and complicated legal battles. Now, "global interferences" may be even more complicated and costly, involving translating records, taking depositions in foreign languages, etc.

In addition to the changes brought about by NAFTA and GATT, other changes in U.S. patent laws are under consideration. One of the major deterrents to full U.S. acceptance of the harmonization has been its unwillingness to adopt a "first-to-file" system. No real progress on this subject has been made. When it was proposed to alter the U.S. system, a strong group of associations, companies, and individuals were in favor of such a change. But, as time has passed, Congress was confronted with the efforts of groups of independent inventors who have promoted claims that it will harm this group. It appears that many people in this country feel that the "first to invent" is the only way the small inventor can be protected from the "large corporation." Whether this is true is yet to be determined.

Although there have been some steps toward harmonization, a complete harmonization treaty is still quite remote. No doubt the patent community, industry, and the legislatures will eventually find a middle ground and proceed. But no one will even offer a guess as to when this will occur.

# Appendix

The following forms used in patent applications and prosecution have been prepared by the U.S. Patent and Trademark Office and are available from the PTO, as well as on the Internet. The form number is printed on the first line in the upper right-hand corner of the page, and also contains a statement about approval date, but does not indicate a date that the use of the form is invalid.

| Form Number | Title |
|---|---|
| PTO/SB/01 | Declaration for Utility or Design Patent Application (2 pages) |
| PTO/SB/02A | Declaration for Additional Inventors |
| PTO/SB/02B | Declaration — Supplemental Priority Data Sheet |
| PTO/SB/02C | Declaration — Registered Practitioner Information |
| PTO/SB/05 | Utility Patent Application Transmittal |
| PTO/SB/09 | Verified Statement Claiming Small Entity Status — Independent Inventor |
| PTO/SB/10 | Statement Claiming Small Entity Status — Small Business |
| PTO/SB/11 | Statement Claiming Small Entity Status — Nonprofit Organization |
| PTO/SB/12 | Statement by a Non-inventor supporting a claim by another for Small Entity Status |
| PTO/SB15 | Assignment of Application |
| PTO/SB/17 | Fee Transmittal |
| PTO/SB/21 | Transmittal Form |
| PTO/SB/16 | Provisional Application for Patent, Cover Sheet (2 pages) |

Please type a plus sign (+) inside this box →

PTO/SB/01 (12-97)
Approved for use through 9/30/00. OMB 0651-0032
Patent and Trademark Office; U.S. DEPARTMENT OF COMMERCE
Under the Paperwork Reduction Act of 1995, no persons are required to respond to a collection of information unless it contains a valid OMB control number.

| DECLARATION FOR UTILITY OR DESIGN PATENT APPLICATION (37 CFR 1.63) | |
|---|---|
| **Attorney Docket Number** | |
| **First Named Inventor** | |
| *COMPLETE IF KNOWN* | |
| Application Number | / |
| Filing Date | |
| Group Art Unit | |
| Examiner Name | |

☐ Declaration Submitted with Initial Filing   **OR**   ☐ Declaration Submitted after Initial Filing (surcharge (37 CFR 1.16 (e)) required)

---

**As a below named inventor, I hereby declare that:**

My residence, post office address, and citizenship are as stated below next to my name.

I believe I am the original, first and sole inventor (if only one name is listed below) or an original, first and joint inventor (if plural names are listed below) of the subject matter which is claimed and for which a patent is sought on the invention entitled:

the specification of which     *(Title of the Invention)*

☐ is attached hereto
OR
☐ was filed on (MM/DD/YYYY) [ ] as United States Application Number or PCT International

Application Number [ ] and was amended on (MM/DD/YYYY) [ ] (if applicable).

I hereby state that I have reviewed and understand the contents of the above identified specification, including the claims, as amended by any amendment specifically referred to above.

I acknowledge the duty to disclose information which is material to patentability as defined in 37 CFR 1.56.

I hereby claim foreign priority benefits under 35 U.S.C. 119(a)-(d) or 365(b) of any foreign application(s) for patent or inventor's certificate, or 365(a) of any PCT international application which designated at least one country other than the United States of America, listed below and have also identified below, by checking the box, any foreign application for patent or inventor's certificate, or of any PCT international application having a filing date before that of the application on which priority is claimed.

| Prior Foreign Application Number(s) | Country | Foreign Filing Date (MM/DD/YYYY) | Priority Not Claimed | Certified Copy Attached? | |
|---|---|---|---|---|---|
| | | | | **YES** | **NO** |
| | | | ☐ | ☐ | ☐ |
| | | | ☐ | ☐ | ☐ |
| | | | ☐ | ☐ | ☐ |
| | | | ☐ | ☐ | ☐ |

☐ Additional foreign application numbers are listed on a supplemental priority data sheet PTO/SB/02B attached hereto:

I hereby claim the benefit under 35 U.S.C. 119(e) of any United States provisional application(s) listed below.

| Application Number(s) | Filing Date (MM/DD/YYYY) | |
|---|---|---|
| | | ☐ Additional provisional application numbers are listed on a supplemental priority data sheet PTO/SB/02B attached hereto. |

[Page 1 of 2]

Burden Hour Statement: This form is estimated to take 0.4 hours to complete. Time will vary depending upon the needs of the individual case. Any comments on the amount of time you are required to complete this form should be sent to the Chief Information Officer, Patent and Trademark Office, Washington, DC 20231. DO NOT SEND FEES OR COMPLETED FORMS TO THIS ADDRESS. SEND TO: Assistant Commissioner for Patents, Washington, DC 20231.

Please type a plus sign (+) inside this box → ☐

PTO/SB/01 (12-97)
Approved for use through 9/30/00. OMB 0651-0032
Patent and Trademark Office; U.S. DEPARTMENT OF COMMERCE
Under the Paperwork Reduction Act of 1995, no persons are required to respond to a collection of information unless it contains a valid OMB control number.

# DECLARATION — Utility or Design Patent Application

I hereby claim the benefit under 35 U.S.C. 120 of any United States application(s), or 365(c) of any PCT international application designating the United States of America, listed below and, insofar as the subject matter of each of the claims of this application is not disclosed in the prior United States or PCT International application in the manner provided by the first paragraph of 35 U.S.C. 112, I acknowledge the duty to disclose information which is material to patentability as defined in 37 CFR 1.56 which became available between the filing date of the prior application and the national or PCT international filing date of this application.

| U.S. Parent Application or PCT Parent Number | Parent Filing Date (MM/DD/YYYY) | Parent Patent Number (if applicable) |
|---|---|---|
| | | |
| | | |

☐ Additional U.S. or PCT international application numbers are listed on a supplemental priority data sheet PTO/SB/02B attached hereto.

As a named inventor, I hereby appoint the following registered practitioner(s) to prosecute this application and to transact all business in the Patent and Trademark Office connected therewith: ☐ Customer Number [ ] ⟶ Place Customer Number Bar Code Label here

OR

☐ Registered practitioner(s) name/registration number listed below

| Name | Registration Number | Name | Registration Number |
|---|---|---|---|
| | | | |
| | | | |

☐ Additional registered practitioner(s) named on supplemental Registered Practitioner Information sheet PTO/SB/02C attached hereto.

Direct all correspondence to: ☐ Customer Number or Bar Code Label [ ] OR ☐ Correspondence address below

| Name | |
|---|---|
| Address | |
| Address | |

| City | | State | | ZIP | |
|---|---|---|---|---|---|
| Country | | Telephone | | Fax | |

I hereby declare that all statements made herein of my own knowledge are true and that all statements made on information and belief are believed to be true; and further that these statements were made with the knowledge that willful false statements and the like so made are punishable by fine or imprisonment, or both, under 18 U.S.C. 1001 and that such willful false statements may jeopardize the validity of the application or any patent issued thereon.

| Name of Sole or First Inventor: | | ☐ A petition has been filed for this unsigned inventor |
|---|---|---|
| Given Name (first and middle [if any]) | | Family Name or Surname |
| | | |

| Inventor s Signature | | | Date | |
|---|---|---|---|---|
| Residence: City | | State | Country | Citizenship |
| Post Office Address | | | | |
| Post Office Address | | | | |
| City | State | ZIP | Country | |

☐ Additional inventors are being named on the ____ supplemental Additional Inventor(s) sheet(s) PTO/SB/02A attached hereto

[Page 2 of 2]

Please type a plus sign (+) inside this box ➔ [ ]

## DECLARATION

**ADDITIONAL INVENTOR(S)**
**Supplemental Sheet**
Page ___ of ___

**Name of Additional Joint Inventor, if any:**    ☐ A petition has been filed for this unsigned inventor

| Given Name (first and middle [if any]) | Family Name or Surname |
|---|---|
| | |

| Inventor's Signature | | Date | |
|---|---|---|---|
| Residence: City | State | Country | Citizenship |
| Post Office Address | | | |
| Post Office Address | | | |
| City | State | ZIP | Country |

**Name of Additional Joint Inventor, if any:**    ☐ A petition has been filed for this unsigned inventor

| Given Name (first and middle [if any]) | Family Name or Surname |
|---|---|
| | |

| Inventor's Signature | | Date | |
|---|---|---|---|
| Residence: City | State | Country | Citizenship |
| Post Office Address | | | |
| Post Office Address | | | |
| City | State | ZIP | Country |

**Name of Additional Joint Inventor, if any:**    ☐ A petition has been filed for this unsigned inventor

| Given Name (first and middle [if any]) | Family Name or Surname |
|---|---|
| | |

| Inventor's Signature | | Date | |
|---|---|---|---|
| Residence: City | State | Country | Citizenship |
| Post Office Address | | | |
| Post Office Address | | | |
| City | State | ZIP | Country |

Please type a plus sign (+) inside this box → ☐

PTO/SB/02B (3-97)
Approved for use through 9/30/98.  OMB 0651-0032
Patent and Trademark Office; U.S. DEPARTMENT OF COMMERCE
Under the Paperwork Reduction Act of 1995, no persons are required to respond to a collection of information unless it contains a
valid OMB control number.

# DECLARATION –– Supplemental Priority Data Sheet

Additional foreign applications:

| Prior Foreign Application Number(s) | Country | Foreign Filing Date (MM/DD/YYYY) | Priority Not Claimed | Certified Copy Attached? YES | NO |
|---|---|---|---|---|---|
| | | | ☐ | ☐ | ☐ |
| | | | ☐ | ☐ | ☐ |
| | | | ☐ | ☐ | ☐ |
| | | | ☐ | ☐ | ☐ |
| | | | ☐ | ☐ | ☐ |
| | | | ☐ | ☐ | ☐ |
| | | | ☐ | ☐ | ☐ |
| | | | ☐ | ☐ | ☐ |
| | | | ☐ | ☐ | ☐ |
| | | | ☐ | ☐ | ☐ |
| | | | ☐ | ☐ | ☐ |

Additional provisional applications:

| Application Number | Filing Date (MM/DD/YYYY) |
|---|---|
| | |

Additional U.S. applications:

| U.S. Parent Application Number | PCT Parent Number | Parent Filing Date (MM/DD/YYYY) | Parent Patent Number (if applicable) |
|---|---|---|---|
| | | | |

Burden Hour Statement: This form is estimated to take 0.4 hours to complete. Time will vary depending upon the needs of the individual case. Any comments on the amount of time  you are required to complete this form should be sent to the Chief Information Officer, Patent and Trademark Office, Washington, DC 20231.  DO NOT SEND FEES OR COMPLETED  FORMS TO THIS ADDRESS. SEND TO: Assistant Commissioner for Patents, Washington, DC 20231.

## DECLARATION

### REGISTERED PRACTITIONER INFORMATION
(Supplemental Sheet)

| Name | Registration Number | Name | Registration Number |
|---|---|---|---|
|  |  |  |  |

Please type a plus sign (+) inside this box → ☐

PTO/SB/05 (4/98)
Approved for use through 09/30/2000. OMB 0651-0032
Patent and Trademark Office: U.S. DEPARTMENT OF COMMERCE
Under the Paperwork Reduction Act of 1995, no persons are required to respond to a collection of information unless it displays a valid OMB control number.

## UTILITY
## PATENT APPLICATION
## TRANSMITTAL
*(Only for new nonprovisional applications under 37 C.F.R. § 1.53(b))*

Attorney Docket No.

First Inventor or Application Identifier

Title

Express Mail Label No.

### APPLICATION ELEMENTS
*See MPEP chapter 600 concerning utility patent application contents.*

ADDRESS TO: Assistant Commissioner for Patents
Box Patent Application
Washington, DC 20231

1. ☐ * Fee Transmittal Form *(e.g., PTO/SB/17)*
   *(Submit an original and a duplicate for fee processing)*

2. ☐ Specification    [Total Pages ☐ ]
   *(preferred arrangement set forth below)*
   - Descriptive title of the Invention
   - Cross References to Related Applications
   - Statement Regarding Fed sponsored R & D
   - Reference to Microfiche Appendix
   - Background of the Invention
   - Brief Summary of the Invention
   - Brief Description of the Drawings *(if filed)*
   - Detailed Description
   - Claim(s)
   - Abstract of the Disclosure

3. ☐ Drawing(s) *(35 U.S.C. 113)*    [Total Sheets ☐ ]

4. Oath or Declaration    [Total Pages ☐ ]
   a. ☐ Newly executed (original or copy)
   b. ☐ Copy from a prior application (37 C.F.R. § 1.63(d))
      *(for continuation/divisional with Box 16 completed)*
      i. ☐ DELETION OF INVENTOR(S)
         Signed statement attached deleting
         inventor(s) named in the prior application,
         see 37 C.F.R. §§ 1.63(d)(2) and 1.33(b).

*NOTE FOR ITEMS 1 & 13: IN ORDER TO BE ENTITLED TO PAY SMALL ENTITY FEES, A SMALL ENTITY STATEMENT IS REQUIRED (37 C.F.R. § 1.27), EXCEPT IF ONE FILED IN A PRIOR APPLICATION IS RELIED UPON (37 C.F.R. § 1.28).*

5. ☐ Microfiche Computer Program *(Appendix)*

6. Nucleotide and/or Amino Acid Sequence Submission
   *(if applicable, all necessary)*
   a. ☐ Computer Readable Copy
   b. ☐ Paper Copy (identical to computer copy)
   c. ☐ Statement verifying identity of above copies

### ACCOMPANYING APPLICATION PARTS

7. ☐ Assignment Papers (cover sheet & document(s))

8. ☐ 37 C.F.R.§3.73(b) Statement ☐ Power of
   *(when there is an assignee)* ☐ Attorney

9. ☐ English Translation Document *(if applicable)*

10. ☐ Information Disclosure ☐ Copies of IDS
    Statement (IDS)/PTO-1449 ☐ Citations

11. ☐ Preliminary Amendment

12. ☐ Return Receipt Postcard (MPEP 503)
    *(Should be specifically itemized)*

13. ☐ * Small Entity ☐ Statement filed in prior application,
    Statement(s) ☐ Status still proper and desired
    *(PTO/SB/09-12)*

14. ☐ Certified Copy of Priority Document(s)
    *(if foreign priority is claimed)*

15. ☐ Other: ....................................................
    ....................................................
    ....................................................

16. If a CONTINUING APPLICATION, check appropriate box, and supply the requisite information below and in a preliminary amendment:
    ☐ Continuation ☐ Divisional ☐ Continuation-in-part (CIP) of prior application No: _____/_____

    Prior application information:    Examiner _____    Group / Art Unit: _____

    **For CONTINUATION or DIVISIONAL APPS only:** The entire disclosure of the prior application, from which an oath or declaration is supplied under Box 4b, is considered a part of the disclosure of the accompanying continuation or divisional application and is hereby incorporated by reference. The incorporation can only be relied upon when a portion has been inadvertently omitted from the submitted application parts.

### 17. CORRESPONDENCE ADDRESS

☐ Customer Number or Bar Code Label
*(Insert Customer No. or Attach bar code label here)*

or ☐ Correspondence address below

| | |
|---|---|
| Name | |
| Address | |

| City | | State | | Zip Code | |
|---|---|---|---|---|---|
| Country | | Telephone | | Fax | |

| Name (Print/Type) | Registration No. (Attorney/Agent) |
|---|---|
| Signature | Date |

Burden Hour Statement: This form is estimated to take 0.2 hours to complete. Time will vary depending upon the needs of the individual case. Any comments on the amount of time you are required to complete this form should be sent to the Chief Information Officer, Patent and Trademark Office, Washington, DC 20231. DO NOT SEND FEES OR COMPLETED FORMS TO THIS ADDRESS. SEND TO: Assistant Commissioner for Patents, Box Patent Application, Washington, DC 20231.

PTO/SB/09 (12-97)
Approved for use through 9/30/00. OMB 0651-0031
Patent and Trademark Office; U.S. DEPARTMENT OF COMMERCE
Under the Paperwork Reduction Act of 1995, no persons are required to respond to a collection of information unless it displays a valid OMB control number.

| STATEMENT CLAIMING SMALL ENTITY STATUS (37 CFR 1.9(f) & 1.27(b))--INDEPENDENT INVENTOR | Docket Number (Optional) |
|---|---|

Applicant, Patentee, or Identifier: _____

Application or Patent No.: _____

Filed or Issued: _____

Title: _____

As a below named inventor, I hereby state that I qualify as an independent inventor as defined in 37 CFR 1.9(c) for purposes of paying reduced fees to the Patent and Trademark Office described in:

☐ the specification filed herewith with title as listed above.

☐ the application identified above.

☐ the patent identified above.

I have not assigned, granted, conveyed, or licensed, and am under no obligation under contract or law to assign, grant, convey, or license, any rights in the invention to any person who would not qualify as an independent inventor under 37 CFR 1.9(c) if that person had made the invention, or to any concern which would not qualify as a small business concern under 37 CFR 1.9(d) or a nonprofit organization under 37 CFR 1.9(e).

Each person, concern, or organization to which I have assigned, granted, conveyed, or licensed or am under an obligation under contract or law to assign, grant, convey, or license any rights in the invention is listed below:

☐ No such person, concern, or organization exists.

☐ Each such person, concern, or organization is listed below.

Separate statements are required from each named person, concern, or organization having rights to the invention stating their status as small entities. (37 CFR 1.27)

I acknowledge the duty to file, in this application or patent, notification of any change in status resulting in loss of entitlement to small entity status prior to paying, or at the time of paying, the earliest of the issue fee or any maintenance fee due after the date on which status as a small entity is no longer appropriate. (37 CFR 1.28(b))

| NAME OF INVENTOR | NAME OF INVENTOR | NAME OF INVENTOR |
|---|---|---|
| Signature of inventor | Signature of inventor | Signature of inventor |
| Date | Date | Date |

Burden Hour Statement: This form is estimated to take 0.2 hours to complete. Time will vary depending upon the needs of the individual case. Any comments on the amount of time you are required to complete this form should be sent to the Chief Information Officer, Patent and Trademark Office, Washington, DC 20231. DO NOT SEND FEES OR COMPLETED FORMS TO THIS ADDRESS. SEND TO: Assistant Commissioner for Patents, Washington, DC 20231.

PTO/SB/10 (12-97)
Approved for use through 9/30/00 OMB 0651-0031
Patent and Trademark Office; U.S. DEPARTMENT OF COMMERCE
Under the Paperwork Reduction Act of 1995, no persons are required to respond to a collection of information unless it displays a valid OMB control number

## STATEMENT CLAIMING SMALL ENTITY STATUS
## (37 CFR 1.9(f) & 1.27(c))–SMALL BUSINESS CONCERN

Docket Number (Optional)

Applicant, Patentee, or Identifier: _____

Application or Patent No.: _____

Filed or Issued: _____

Title: _____

I hereby state that I am

☐ the owner of the small business concern identified below:

☐ an official of the small business concern empowered to act on behalf of the concern identified below:

NAME OF SMALL BUSINESS CONCERN _____

ADDRESS OF SMALL BUSINESS CONCERN _____

I hereby state that the above identified small business concern qualifies as a small business concern as defined in 13 CFR Part 121 for purposes of paying reduced fees to the United States Patent and Trademark Office, in that the number of employees of the concern, including those of its affiliates, does not exceed 500 persons. For purposes of this statement, (1) the number of employees of the business concern is the average over the previous fiscal year of the concern of the persons employed on a full-time, part-time, or temporary basis during each of the pay periods of the fiscal year, and (2) concerns are affiliates of each other when either, directly or indirectly, one concern controls or has the power to control the other, or a third party or parties controls or has the power to control both.

I hereby state that rights under contract or law have been conveyed to and remain with the small business concern identified above with regard to the invention described in:

☐ the specification filed herewith with title as listed above.

☐ the application identified above.

☐ the patent identified above.

If the rights held by the above identified small business concern are not exclusive, each individual, concern, or organization having rights in the invention must file separate statements as to their status as small entities, and no rights to the invention are held by any person, other than the inventor, who would not qualify as an independent inventor under 37 CFR 1.9(c) if that person made the invention, or by any concern which would not qualify as a small business concern under 37 CFR 1.9(d), or a nonprofit organization under 37 CFR 1.9(e).

☐ Each person, concern, or organization having any rights in the invention is listed below:

☐ no such person, concern, or organization exists.

☐ each such person, concern, or organization is listed below.

Separate statements are required from each named person, concern or organization having rights to the invention stating their status as small entities. (37 CFR 1.27)

I acknowledge the duty to file, in this application or patent, notification of any change in status resulting in loss of entitlement to small entity status prior to paying, or at the time of paying, the earliest of the issue fee or any maintenance fee due after the date on which status as a small entity is no longer appropriate. (37 CFR 1.28(b))

NAME OF PERSON SIGNING _____

TITLE OF PERSON IF OTHER THAN OWNER _____

ADDRESS OF PERSON SIGNING _____

SIGNATURE _____ DATE _____

Burden Hour Statement: This form is estimated to take 0.2 hours to complete. Time will vary depending upon the needs of the individual case. Any comments on the amount of time you are required to complete this form should be sent to the Chief Information Officer, Patent and Trademark Office, Washington, DC 20231. DO NOT SEND FEES OR COMPLETED FORMS TO THIS ADDRESS. SEND TO: Assistant Commissioner for Patents, Washington, DC 20231.

PTO/SB/11 (12-97)
Approved for use through 9/30/00. OMB 0651-0031
Patent and Trademark Office; U.S. DEPARTMENT OF COMMERCE
Under the Paperwork Reduction Act of 1995, no persons are required to respond to a collection of information unless it displays a valid OMB control number.

| STATEMENT CLAIMING SMALL ENTITY STATUS (37 CFR 1.9(f) & 1.27(d))--NONPROFIT ORGANIZATION | Docket Number (Optional) |
|---|---|

Applicant, Patentee, or Identifier: _____

Application or Patent No.: _____

Filed or Issued: _____

Title: _____

I hereby state that I am an official empowered to act on behalf of the nonprofit organization identified below:

NAME OF NONPROFIT ORGANIZATION _____

ADDRESS OF NONPROFIT ORGANIZATION _____

TYPE OF NONPROFIT ORGANIZATION:

☐ UNIVERSITY OR OTHER INSTITUTION OF HIGHER EDUCATION

☐ TAX EXEMPT UNDER INTERNAL REVENUE SERVICE CODE (26 U.S.C. 501(a) and 501(c)(3))

☐ NONPROFIT SCIENTIFIC OR EDUCATIONAL UNDER STATUTE OF STATE OF THE UNITED STATES OF AMERICA
(NAME OF STATE _____ )
(CITATION OF STATUTE _____ )

☐ WOULD QUALIFY AS TAX EXEMPT UNDER INTERNAL REVENUE SERVICE CODE (26 U.S.C. 501(a) and 501(c)(3)) IF LOCATED IN THE UNITED STATES OF AMERICA

☐ WOULD QUALIFY AS NONPROFIT SCIENTIFIC OR EDUCATIONAL UNDER STATUTE OF STATE OF THE UNITED STATES OF AMERICA IF LOCATED IN THE UNITED STATES OF AMERICA
(NAME OF STATE _____ )
(CITATION OF STATUTE _____ )

I hereby state that the nonprofit organization identified above qualifies as a nonprofit organization as defined in 37 CFR 1.9(e) for purposes of paying reduced fees to the United States Patent and Trademark Office regarding the invention described in:

☐ the specification filed herewith with title as listed above.
☐ the application identified above.
☐ the patent identified above.

I hereby state that rights under contract or law have been conveyed to and remain with the nonprofit organization regarding the above identified invention. If the rights held by the nonprofit organization are not exclusive, each individual, concern, or organization having rights in the invention must file separate statements as to their status as small entities and that no rights to the invention are held by any person, other than the inventor, who would not qualify as an independent inventor under 37 CFR 1.9(c) if that person made the invention, or by any concern which would not qualify as a small business concern under 37 CFR 1.9(d) or a nonprofit organization under 37 CFR 1.9(e).

Each person, concern, or organization having any rights in the invention is listed below:

☐ no such person, concern, or organization exists.
☐ each such person, concern, or organization is listed below.

I acknowledge the duty to file, in this application or patent, notification of any change in status resulting in loss of entitlement to small entity status prior to paying, or at the time of paying, the earliest of the issue fee or any maintenance fee due after the date on which status as a small entity is no longer appropriate. (37 CFR 1.28(b))

NAME OF PERSON SIGNING _____

TITLE IN ORGANIZATION OF PERSON SIGNING _____

ADDRESS OF PERSON SIGNING _____

SIGNATURE _____ DATE _____

Burden Hour Statement: This form is estimated to take 0.2 hours to complete. Time will vary depending upon the needs of the individual case. Any comments on the amount of time you are required to complete this form should be sent to the Chief Information Officer, Patent and Trademark Office, Washington, DC 20231. DO NOT SEND FEES OR COMPLETED FORMS TO THIS ADDRESS. SEND TO: Assistant Commissioner for Patents, Washington, DC 20231.

PTO/SB/12 (12-97)
Approved for use through 9/30/00. OMB 0651-0031
Patent and Trademark Office; U.S. DEPARTMENT OF COMMERCE
Under the Paperwork Reduction Act of 1995, no persons are required to respond to a collection of information unless it displays a valid OMB control number.

## STATEMENT BY A NON-INVENTOR SUPPORTING
## A CLAIM BY ANOTHER FOR SMALL ENTITY STATUS

Docket Number (Optional)

Applicant, Patentee, or Identifier: _____

Application or Patent No.: _____

Filed or Issued: _____

Title: _____

I hereby state that I am making this statement to support a claim by _____ for small entity status for purposes of paying reduced fees to the United States Patent and Trademark Office, regarding the invention described in:

☐ the specification filed herewith with title as listed above.
☐ the application identified above.
☐ the patent identified above.

I hereby state that I would qualify as an independent inventor as defined in 37 CFR 1.9(c) for purposes of paying fees to the United States Patent and Trademark Office, if I had made the above identified invention.

I have not assigned, granted, conveyed or licensed and am under no obligation under contract or law to assign, grant, convey or license, any rights in the invention to any person who would not qualify as an independent inventor under 37 CFR 1.9(c) if that person had made the invention, or to any concern which would not qualify as a small business concern under 37 CFR 1.9(d) or a nonprofit organization under 37 CFR 1.9(e). Note: Separate statements are required from each person, concern or organization having rights to the invention to their status as small entities. (37 CFR 1.27)

Each person, concern, or organization to which I have assigned, granted, conveyed, or licensed or am under an obligation under contract or law to assign, grant, convey, or license any rights in the invention is listed below:

☐ no such person, concern, or organization exists.
☐ each such person, concern, or organization is listed below.

I acknowledge the duty to file, in this application or patent, notification of any change in status resulting in loss of entitlement to small entity status prior to paying, or at the time of paying, the earliest of the issue fee or any maintenance fee due after the date on which status as a small entity is no longer appropriate. (37 CFR 1.28(b))

NAME OF PERSON SIGNING _____

TITLE IN ORGANIZATION OF PERSON SIGNING _____

ADDRESS OF PERSON SIGNING _____

SIGNATURE _____  DATE _____

Burden Hour Statement: This form is estimated to take 0.2 hours to complete. Time will vary depending upon the needs of the individual case. Any comments on the amount of time you are required to complete this form should be sent to the Chief Information Officer, Patent and Trademark Office, Washington, DC 20231. DO NOT SEND FEES OR COMPLETED FORMS TO THIS ADDRESS. SEND TO: Assistant Commissioner for Patents, Washington, DC 20231.

PTO/SB/15 (8-96)
Approved for use through 9/30/98.  OMB 0651-0027
Patent and Trademark Office; U.S. DEPARTMENT OF COMMERCE
Under the Paperwork Reduction Act of 1995, no persons are required to respond to a collection of information unless it displays a valid OMB control number

| ASSIGNMENT OF APPLICATION | Docket Number (Optional) |
|---|---|

Whereas, I, _____ of _____ , hereafter

referred to as applicant, have invented certain new and useful improvements in _____

_____

☐ for which an application for a United States Patent was filed on _____ ,
Application Number _____/_____ .

☐ for which an application for a United States Patent was executed on _____ , and

Whereas, _____ of _____ herein referred to

"assignee" whose post office address is _____ is de-

sirous of acquiring the entire right, title and interest in the same;

Now, therefore, in consideration of the sum of _____ dollars ($_____ ), the receipt whereof is ac-

knowledged, and other good and valuable consideration, I, the applicant, by these presents do sell, assign

and transfer unto said assignee the full and exclusive right to the said invention in the United States and the

entire right, title and interest in and to any and all Patents which may be granted therefor in the United States,

I hereby authorize and request the Commissioner of Patents and Trademarks to issue said United States

Patent to said assignee, of the entire right, title, and interest in and to the same, for his sole use and behoof;

and for the use and behoof of his legal representatives, to the full end of the term for which said Patent may

be granted, as fully and entirely as the same would have been held by me had this assignment and sale not

been made.

Executed this _____ day of _____ , 19_____ ,

at _____ .

_____
(Signature)

State of _____ )   SS:
County of _____ )
Before me personally appeared said _____
and acknowledged the foregoing instrument to be his free act and deed this _____
day of _____ , 19____ .

Seal

_____
(Notary Public)

Burden Hour Statement:  This form is estimated to take 0.1 hours to complete.  Time will vary depending upon the needs of the individual case.  Any comments on the amount of time you are required to complete this form should be sent to the Chief Information Officer, Patent and Trademark Office, Washington, DC 20231.  DO NOT SEND FEES OR COMPLETED FORMS TO THIS ADDRESS.  SEND TO:  Commissioner of Patents and Trademarks, Washington, DC 20231.

PTO/SB/17 (2/98)
Approved for use through 9/30/2000. OMB 0651-0032
Patent and Trademark Office: U.S. DEPARTMENT OF COMMERCE
Under the Paperwork Reduction Act of 1995, no persons are required to respond to a collection of information unless it displays a valid OMB control number.

# FEE TRANSMITTAL

*Patent fees are subject to annual revision on October 1.*
*These are the fees effective October 1, 1997.*
*Small Entity payments must be supported by a small entity statement,*
*otherwise large entity fees must be paid. See Forms PTO/SB/09-12.*
*See 37 C.F.R. §§ 1.27 and 1.28.*

**TOTAL AMOUNT OF PAYMENT     ($)**

**Complete if Known**

| | |
|---|---|
| Application Number | |
| Filing Date | |
| First Named Inventor | |
| Examiner Name | |
| Group / Art Unit | |
| Attorney Docket No. | |

## METHOD OF PAYMENT (check one)

1. ☐ The Commissioner is hereby authorized to charge indicated fees and credit any over payments to:

Deposit Account Number

Deposit Account Name

☐ Charge Any Additional Fee Required Under 37 C.F.R. §§ 1.16 and 1.17

☐ Charge the Issue Fee Set in 37 C.F.R. § 1.18 at the Mailing of the Notice of Allowance

2. ☐ **Payment Enclosed:**
   ☐ Check   ☐ Money Order   ☐ Other

## FEE CALCULATION

### 1. BASIC FILING FEE

| Large Entity Fee Code | Fee ($) | Small Entity Fee Code | Fee ($) | Fee Description | Fee Paid |
|---|---|---|---|---|---|
| 101 | 790 | 201 | 395 | Utility filing fee | |
| 106 | 330 | 206 | 165 | Design filing fee | |
| 107 | 540 | 207 | 270 | Plant filing fee | |
| 108 | 790 | 208 | 395 | Reissue filing fee | |
| 114 | 150 | 214 | 75 | Provisional filing fee | |

**SUBTOTAL (1)   ($)**

### 2. EXTRA CLAIM FEES

| | Extra Claims | Fee from below | Fee Paid |
|---|---|---|---|
| Total Claims | ☐ -20**= | X ☐ = | ☐ |
| Independent Claims | ☐ - 3**= | X ☐ = | ☐ |
| Multiple Dependent | ☐ | = | ☐ |

**or number previously paid, if greater; For Reissues, see below

| Large Entity Fee Code | Fee ($) | Small Entity Fee Code | Fee ($) | Fee Description |
|---|---|---|---|---|
| 103 | 22 | 203 | 11 | Claims in excess of 20 |
| 102 | 82 | 202 | 41 | Independent claims in excess of 3 |
| 104 | 270 | 204 | 135 | Multiple dependent claim, if not paid |
| 109 | 82 | 209 | 41 | ** Reissue independent claims over original patent |
| 110 | 22 | 210 | 11 | ** Reissue claims in excess of 20 and over original patent |

**SUBTOTAL (2)   ($)**

## FEE CALCULATION (continued)

### 3. ADDITIONAL FEES

| Large Entity Fee Code | Fee ($) | Small Entity Fee Code | Fee ($) | Fee Description | Fee Paid |
|---|---|---|---|---|---|
| 105 | 130 | 205 | 65 | Surcharge - late filing fee or oath | |
| 127 | 50 | 227 | 25 | Surcharge - late provisional filing fee or cover sheet. | |
| 139 | 130 | 139 | 130 | Non-English specification | |
| 147 | 2,520 | 147 | 2,520 | For filing a request for reexamination | |
| 112 | 920* | 112 | 920* | Requesting publication of SIR prior to Examiner action | |
| 113 | 1,840* | 113 | 1,840* | Requesting publication of SIR after Examiner action | |
| 115 | 110 | 215 | 55 | Extension for reply within first month | |
| 116 | 400 | 216 | 200 | Extension for reply within second month | |
| 117 | 950 | 217 | 475 | Extension for reply within third month | |
| 118 | 1,510 | 218 | 755 | Extension for reply within fourth month | |
| 128 | 2,060 | 228 | 1,030 | Extension for reply within fifth month | |
| 119 | 310 | 219 | 155 | Notice of Appeal | |
| 120 | 310 | 220 | 155 | Filing a brief in support of an appeal | |
| 121 | 270 | 221 | 135 | Request for oral hearing | |
| 138 | 1,510 | 138 | 1,510 | Petition to institute a public use proceeding | |
| 140 | 110 | 240 | 55 | Petition to revive - unavoidable | |
| 141 | 1,320 | 241 | 660 | Petition to revive - unintentional | |
| 142 | 1,320 | 242 | 660 | Utility issue fee (or reissue) | |
| 143 | 450 | 243 | 225 | Design issue fee | |
| 144 | 670 | 244 | 335 | Plant issue fee | |
| 122 | 130 | 122 | 130 | Petitions to the Commissioner | |
| 123 | 50 | 123 | 50 | Petitions related to provisional applications | |
| 126 | 240 | 126 | 240 | Submission of Information Disclosure Stmt | |
| 581 | 40 | 581 | 40 | Recording each patent assignment per property (times number of properties) | |
| 146 | 790 | 246 | 395 | Filing a submission after final rejection (37 CFR 1.129(a)) | |
| 149 | 790 | 249 | 395 | For each additional invention to be examined (37 CFR 1.129(b)) | |

Other fee (specify) _____

Other fee (specify) _____

* Reduced by Basic Filing Fee Paid

**SUBTOTAL (3)   ($)**

| SUBMITTED BY | | Complete (if applicable) | |
|---|---|---|---|
| Typed or Printed Name | | Reg. Number | |
| Signature | Date | Deposit Account User ID | |

Burden Hour Statement: This form is estimated to take 0.2 hours to complete. Time will vary depending upon the needs of the individual case. Any comments on the amount of time you are required to complete this form should be sent to the Chief Information Officer, Patent and Trademark Office, Washington, DC 20231. DO NOT SEND FEES OR COMPLETED FORMS TO THIS ADDRESS. SEND TO: Assistant Commissioner for Patents, Washington, DC 20231.

PTO/SB/21 (12-97)
Approved for use through 9/30/00. OMB 0651-0031
Patent and Trademark Office: U.S. DEPARTMENT OF COMMERCE

Please type a plus sign (+) inside this box → ☐

Under the Paperwork Reduction Act of 1995, no persons are required to respond to a collection of information unless it displays a valid OMB control number.

# TRANSMITTAL FORM

*(to be used for all correspondence after initial filing)*

| | |
|---|---|
| Application Number | |
| Filing Date | |
| First Named Inventor | |
| Group Art Unit | |
| Examiner Name | |
| Total Number of Pages in This Submission | Attorney Docket Number |

## ENCLOSURES *(check all that apply)*

☐ Fee Transmittal Form
    ☐ Fee Attached

☐ Amendment / Response
    ☐ After Final
    ☐ Affidavits/declaration(s)

☐ Extension of Time Request

☐ Express Abandonment Request

☐ Information Disclosure Statement

☐ Certified Copy of Priority Document(s)

☐ Response to Missing Parts/ Incomplete Application
    ☐ Response to Missing Parts under 37 CFR 1.52 or 1.53

☐ Assignment Papers *(for an Application)*

☐ Drawing(s)

☐ Licensing-related Papers

☐ Petition Routing Slip (PTO/SB/69) and Accompanying Petition

☐ To Convert a Provisional Application

☐ Power of Attorney, Revocation Change of Correspondence Address

☐ Terminal Disclaimer

☐ Small Entity Statement

☐ Request for Refund

Remarks

☐ After Allowance Communication to Group

☐ Appeal Communication to Board of Appeals and Interferences

☐ Appeal Communication to Group *(Appeal Notice, Brief, Reply Brief)*

☐ Proprietary Information

☐ Status Letter

☐ Additional Enclosure(s) *(please identify below):*

## SIGNATURE OF APPLICANT, ATTORNEY, OR AGENT

| | |
|---|---|
| Firm or Individual name | |
| Signature | |
| Date | |

## CERTIFICATE OF MAILING

I hereby certify that this correspondence is being deposited with the United States Postal Service as first class mail in an envelope addressed to: Assistant Commissioner for Patents, Washington, D.C. 20231 on this date:

| | |
|---|---|
| Typed or printed name | |
| Signature | Date |

Burden Hour Statement: This form is estimated to take 0.2 hours to complete. Time will vary depending upon the needs of the individual case. Any comments on the amount of time you are required to complete this form should be send to the Chief Information Officer, Patent and Trademark Office, Washington, DC 20231. DO NOT SEND FEES OR COMPLETED FORMS TO THIS ADDRESS. SEND TO: Assistant Commissioner for Patents, Washington, DC 20231.

Please type a plus sign (+) inside this box ⟶ ☐

PTO/SB/16 (2-98)
Approved for use through 01/31/2001. OMB 0651-0037
Patent and Trademark Office; U.S. DEPARTMENT OF COMMERCE
Under the Paperwork Reduction Act of 1995, no persons are required to respond to a collection of information unless it displays a valid OMB control number.

## PROVISIONAL APPLICATION FOR PATENT COVER SHEET
### This is a request for filing a PROVISIONAL APPLICATION FOR PATENT under 37 CFR 1.53 (c).

### INVENTOR(S)

| Given Name (first and middle [if any]) | Family Name or Surname | Residence (City and either State or Foreign Country) |
|---|---|---|
|  |  |  |

☐ Additional inventors are being named on the ___ separately numbered sheets attached hereto

### TITLE OF THE INVENTION (280 characters max)

**CORRESPONDENCE ADDRESS**

Direct all correspondence to:

☐ Customer Number

Type Customer Number here ⟶

Place Customer Number Bar Code Label here

OR

☐ Firm or Individual Name

| Address |  |
|---|---|
| Address |  |
| City |  | State |  | ZIP |
| Country |  | Telephone |  | Fax |

### ENCLOSED APPLICATION PARTS (check all that apply)

☐ Specification *Number of Pages* [   ]   ☐ Small Entity Statement

☐ Drawing(s) *Number of Sheets* [   ]   ☐ Other (specify) [   ]

### METHOD OF PAYMENT OF FILING FEES FOR THIS PROVISIONAL APPLICATION FOR PATENT *(check one)*

☐ A check or money order is enclosed to cover the filing fees

FILING FEE AMOUNT ($)

☐ The Commissioner is hereby authorized to charge filing fees or credit any overpayment to Deposit Account Number:

The invention was made by an agency of the United States Government or under a contract with an agency of the United States Government.
☐ No.
☐ Yes, the name of the U.S. Government agency and the Government contract number are:_____

Respectfully submitted,

Date    /    /

SIGNATURE _____

TYPED or PRINTED NAME _____

REGISTRATION NO. (if appropriate)

Docket Number:

TELEPHONE _____

## USE ONLY FOR FILING A PROVISIONAL APPLICATION FOR PATENT

This collection of information is required by 37 CFR 1.51. The information is used by the public to file (and by the PTO to process) a provisional application. Confidentiality is governed by 35 U.S.C. 122 and 37 CFR 1.14. This collection is estimated to take 8 hours to complete, including gathering, preparing, and submitting the complete provisional application to the PTO. Time will vary depending upon the individual case. Any comments on the amount of time you require to complete this form and/or suggestions for reducing this burden, should be sent to the Chief Information Officer, U.S. Patent and Trademark Office, U.S. Department of Commerce, Washington, D.C., 20231. DO NOT SEND FEES OR COMPLETED FORMS TO THIS ADDRESS. SEND TO: Box Provisional Application, Assistant Commissioner for Patents, Washington, D.C., 20231.

### PROVISIONAL APPLICATION COVER SHEET
### Additional Page

PTO/SB/16 (2-98)
Approved for use through 01/31/2001. OMB 0651-0037
Patent and Trademark Office; U.S. DEPARTMENT OF COMMERCE
Under the Paperwork Reduction Act of 1995, no persons are required to respond to a collection of information unless it displays a valid OMB control number.

| Docket Number | | Type a plus sign (+) inside this box → |
|---|---|---|

**INVENTOR(S)/APPLICANT(S)**

| Given Name (first and middle [if any]) | Family or Surname | Residence (City and either State or Foreign Country) |
|---|---|---|
| | | |

Number ____ of ____

Questions & Answers

Relating To

# Provisional Applications For Patent

United States Patent and Trademark Office
March 1998

1.  Can a patent be issued on a provisional application?

        Ans.:  A patent can be issued on a subsequently filed
nonprovisional application that claims the benefits of a provisional
application filing date.  The provisional application, itself, cannot
mature into a patent.

2.  Can you use a provisional application for a design patent
application?

        Ans.:  No.  Provisional applications are not available for design
inventions.

3.  Can a design application claim priority benefits of a prior
provisional application?

        Ans.:  No, the design sections of the statute preclude the claiming
of priority benefits in design patent applications based on prior
provisional applications.

4.  What are the requirements for filing a provisional application for
patent?

        Ans.:  A written description of the invention, complying with 35
U.S.C. § 112, first paragraph; drawings, complying with 35 U.S.C. § 113;
the filing fee; and a cover sheet.

5.  What information is required on the cover sheet?

    Ans.:  The cover sheet must identify the:

            (1) application as a provisional application;
            (2) name(s) of the inventor(s);
            (3) residence of each named inventor (city & state or city &
                foreign country);
            (4) title of the invention;
            (5) name and registration number of atty./agent, if
                applicable;
            (6) docket number, if applicable;
            (7) correspondence address; and
            (8) name of the U.S. government agency and government
        contract number (if the invention was made by an agency
     of the U.S. government or under contract with an agency
of the U.S. government).

6.  Is there a specific format for the provisional application cover
sheet?

        Ans.:  No.  The PTO has a suggested cover sheet which is available
on request without charge.  However, the cover sheet requirements are
content, not format, requirements.  Applicants may design their own
cover sheets so long as the informational content requirements are
satisfied.

7.  What is the current filing fee for a provisional application?

2

Ans.: The current filing fee for a provisional application may be obtained by contacting the Patent and Trademark Office General Information Services Division at (800) PTO-9199 or (703) 308-HELP. The filing fee for a provisional application is subject to a 50% reduction for a small entity who has properly established status as a small entity at the time the filing fee is paid. The filing fee may be paid by personal check made payable to the "Assistant Commissioner for Patents."

8. Does a small entity applicant have to file a small entity statement to pay the small entity filing fee?

Ans.: Yes.

9. Is any particular format required for a provisional application?

Ans.: No. However, applicants should follow generally the format set forth in 37 CFR 1.77 when applicable.

10. Are there any formal requirements for the specification and drawings in the provisional application?

Ans.: Yes. All provisional applications are optically scanned and stored in a secure electronic data base. In order to permit scanning, all pages of the specification must be legibly written either by typewriter or mechanical printer in permanent dark ink in portrait orientation. The lines of the specification must be 1 ½ or double spaced. Each sheet of specification and drawings must be (1) presented on flexible, strong, smooth, non-shiny, durable, and white paper, (2) written on only one side and (3) either 21.0 cm. by 29.7 cm. (DIN size A4) or 8 ½ by 11 inches. Each sheet of specification must include a top, right side and bottom margin of at least 2.0 cm. (3/4 inch) and a left side margin of at least 2.5 cm. (1 inch). Each sheet of the drawings must include a top and left side margin of at least 2.5 cm. (1 inch), a right side margin of at least 1.5 cm. (5/8 inch), and a bottom margin of at least 1.0 cm. (3/8 inch). Photographs must either be developed on double weight photographic paper or be permanently mounted on Bristol board.

11. Does a provisional application require a claim?

Ans.: No. However, claims may be included in a provisional application.

12. Does a provisional application require an oath or declaration of the inventor?

Ans.: No. However, an oath or declaration may be included in a provisional application.

13. Does a provisional application require a power of attorney?

Ans.: No. However, a power of attorney may facilitate access to a provisional application file.

14. Does the description in a provisional application have to comply

with the first paragraph of 35 U.S.C. § 112?

Ans.: Yes. An applicant is entitled to claim benefit of a provisional application only to the extent that a later claimed invention in a nonprovisional application is described in the provisional application in the manner required by 35 U.S.C. § 112, first paragraph.

15. Does the best mode have to be described in a provisional application?

Ans.: Yes, for the reasons set forth above.

16. Will a filing date receipt be issued in a provisional application?

Ans.: Yes.

17. Is a provisional application a regular national filing for the purpose of the Paris Convention?

Ans.: Yes. Foreign filings must be made within twelve months of the filing date of the provisional application. An applicant must file internationally within twelve months of filing a provisional application if the provisional application filing date is to be relied on.

18. Will the PTO grant foreign filing licenses based upon the filing of a provisional application?

Ans.: Yes. Since a provisional application is a regular national filing, it starts the Paris Convention year in order to file applications in foreign countries and obtain the benefit for a prior filing date in the United States. Because the provisional application will form the basis for foreign filings, it will be screened and a foreign filing license issued based thereon.

19. Can a provisional application claim the benefit of the filing date of another application?

Ans.: No, a provisional application cannot claim the benefit of an earlier filed application, either domestic or foreign.

20. Will an examiner review the content of a provisional application?

Ans.: An examiner will review the content of a provisional application only in those situations where it is necessary to determine if a nonprovisional application or a patent claiming benefits of a prior provisional application is actually entitled to the filing date of the provisional application.

21. Can a provisional application be pending for more than a year?

Ans.: No.

22. Will provisional applications become available to the public?

4

Ans.: By statute, provisional applications are considered abandoned one year after they are filed. Accordingly, they will **not**, simply as a provisional application, be available to the public. However, in most cases, a nonprovisional application will be filed making reference to the provisional application so that once a patent issues on the nonprovisional application making reference to the provisional application, the provisional application would be available to the public in the same way that any patent application on which an issued patent is based is now available to the public.

23. Can an amendment be made to a provisional application?

Ans.: No. No amendment or submission can be made in a provisional application unless it is in response to an Office requirement. In order to add new material to a provisional application, a second provisional application containing the new material must be filed. Note, the second provisional application cannot rely upon the first but a subsequently filed nonprovisional application may rely, separately, on both provisional applications.

24. Can you rely on a plurality of provisional applications in a subsequent filed nonprovisional application?

Ans.: Yes. However, a claim in the nonprovisional application is entitled to the provisional application filing date only to the extent that the subject matter of the claim is supported in a particular provisional application. Also, the nonprovisional application must be filed within 12 months of the filing date of each provisional application and each provisional application must be pending on the filing date of the nonprovisional application.

25. Can a person filing a provisional application use "patent pending" on a product that is marketed?

Ans.: The provisional application is clearly an application for patent that clearly signifies that the inventor has entered the patent system and has taken a first step to obtaining a patent. However, a provisional application will not be pending after a year from its filing date, so unless the inventor has filed another application, the marking would not be appropriate after a year.

26. How should a provisional application be referred to in a subsequently filed nonprovisional application?

Ans.: All provisional applications will be given application numbers starting with a series code "60," then a six digit number, e.g., "60/123,456." This number and the provisional application filing date will serve to identify the provisional application. The reference to the provisional application may read, "This application claims the benefit of U.S. Provisional Application No. 60/_____ filed on _____."

27. What address should be used to file a provisional application by mail?

5

Ans.: All provisional applications and papers relating thereto should be addressed as follows:

> Assistant Commissioner for Patents
> Box Provisional Patent Application
> Washington, D.C. 20231

28. Will a provisional patent applicant be able to manufacture or sell in the provisional year?

Ans.: Yes, subject, of course, to the patent rights of others.

29. Is the one year grace period for filing an application after the first public use or sale in the U.S. or the first printed publication describing the invention measured from the provisional application filing date or the nonprovisional application filing date?

Ans.: To the extent that a nonprovisional application is entitled to the benefits of the filing date of a prior provisional application, the grace period is measured from the provisional application filing date.

30. Can provisional applications be assigned? If so, does the subsequent nonprovisional application have to be separately assigned?

Ans.: Yes, a provisional application can be assigned. Separate assignments for the provisional application and subsequently filed nonprovisional application should be submitted if the subsequently filed nonprovisional application includes subject matter not included in the provisional application.

31. Will the filing of a provisional application affect the order of examination of a nonprovisional application that relies on a provisional application? That is, will a nonprovisional application be taken up for examination based upon the nonprovisional application filing date or the provisional application filing date?

Ans.: No. Since the provisional application filing date does not start the patent term, starting examination in a nonprovisional application based on the provisional application filing date would not be fair relative to those applications that do not rely on a provisional application.

# Glossary

## Words and phrases used in patent terminology

| | |
|---|---|
| **Abandon:** | To lose, by direct or indirect action, rights in a patent or patent application. Direct action such as relinquishing a claim to a patent, or indirectly by failure to respond to an Office Action. |
| **Abstract:** | A required part of a patent application; a brief summary of the patent application. This differs from the Summary of the Invention section of the application. The abstract should enable the reader to determine the character of the subject matter and declare what is new in the art area. |
| **Affidavit:** | A written, sworn statement describing facts and additional information supporting the patent application. |
| **Allowance, Notice of:** | See Notice of Allowance. |
| **Amendment:** | A written statement to the Patent Office to correct or change any part of the application. The amendment is usually submitted in response to an action by the patent examiner. |
| **Annuity:** | An annual fee payable to a foreign country to continue a patent in force. See Maintenance Fees, which are the same type of fee, but not payable annually. |
| **Anticipation:** | A portion of the prior art that is similar enough to negate the novelty of the invention. |
| **Appeal:** | An answer to an examiner's action, directed to the Board of Patent Appeal and Interference, requesting specific action. |
| **Appeal Brief:** | A document that accompanies the Appeal, which sets forth the applicant's argument, the legal basis, discussion, conclusion, and recommendation for action. |
| **Appeal, Notice of:** | A document accompanying the required fee, stating that an appeal will be filed in a specific case. |

| | |
|---|---|
| **Applicant:** | Person or persons applying for a patent, who claim to be the inventor or inventors of the patent application being filed. |
| **Application:** | The formal document comprising the specification, claims, oath, declaration, drawings (if necessary), and filing fee upon which an inventor seeks a patent. |
| **Art:** | An area or field of technology or technical data. |
| **Article of Manufacture:** | A class of patentable subject matter. Any item that can be made by man. |
| **Assignee:** | The person or organization to whom the rights associated with the invention have been transferred. |
| **Assignment:** | The legal document that describes the terms and conditions under which the rights to the invention are transferred to another. |
| **Assignor:** | The person or organization that owns the patent rights and is transferring them to another. |
| **Auslegeschrift:** | The name of a German patent application that is published for inspection and opposition of the public. |
| **Background of Invention:** | A section of the patent application in which the technological area of the invention is described. |
| **Best Mode:** | The best way, known to the inventor, of making or operating the invention which must be presented in the patent application. |
| **Board of Patent Appeals and Interferences:** | A group of senior officials within the Patent Branch of the Patent and Trademark Office that considers the actions brought by an applicant's appeal of an examiner's action. The board also determines, in an interference action, which of two or more inventors was the first to invent and is therefore entitled to a patent. |
| **CAFC:** | The abbreviation for the Court of Appeals for the Federal Circuit, which was created in 1982 to replace the Court of Customs and Patent Appeal (CCPA). The court has jurisdiction in all areas of patents, trademarks, and copyrights. Appeals from PTO actions are taken to this court. |
| **Certificate of Correction:** | A document issued by the Patent and Trademark Office as an addendum to an issued utility patent correcting minor errors in the original patent. |
| **CFR:** | An abbreviation for *Code of Federal Regulations*. Volume 37 includes the rules and regulations for patents, trademarks, and copyrights. |
| **CIP:** | An abbreviation for a Continuation in Part application. See below. |
| **Claim:** | The precise descriptions of the part of the invention for which the inventor is seeking patent protection. The claims define the invention. |

| | |
|---|---|
| **Claim, Dependent:** | A claim including, by reference to another claim, all of its subject matter and containing some further restriction or limitation. |
| **Claim, Independent:** | A claim that does not have any reference to any other claim, and stands by itself. |
| **Claim, Process:** | A claim to the method of the invention. |
| **Claim, Product:** | A claim to the physical form of an invention or to an invention whose form is physical. (A chemical compound as opposed to a method for making the compound.) |
| **Composition of Matter:** | A statutory class of invention in which the physical mixture, or chemically combined substances, is the inventive subject. |
| **Comprising:** | A term used in writing claims, inserted between the preamble and the body of the claim, that expands the claim to include what is specifically stated and everything that could be added. *See* **Consisting of**. |
| **Conception:** | The act of visualizing an invention, complete in all detail. It occurs when a solution to a problem is formulated, not when the problem is realized. |
| **Consisting Essentially of:** | A term used in writing claims that is intermediate in scope between the terms "comprising" and "consisting of." |
| **Consisting of:** | A term used in writing claims that limits the breadth of the claim to what is specifically specified. |
| **Continuation:** | An application filed during the pendency of an earlier filed (also known as Parent) application, by the same applicant, containing the same disclosure as the earlier filed application, but with different claims. |
| **Continuation in Part:** | An application filed during the pendency of an earlier filed (Parent) application by the same inventor, disclosing substantial portions of the subject matter of the earlier filed application and some new subject matter. |
| **Constructive Reduction to Practice:** | An act signified by the filing of a patent application, which is legally equivalent to actual reduction to practice. *See* **Reduction to Practice**. |
| **Copyright:** | A right bestowed by the government to protect the original artistic and literary works of an author. The right is given by the Library of Congress, not the Patent and Trademark Office. |
| **Corroboration:** | Evidence from a person, not an inventor, that demonstrates and supports the inventive act or dates. |
| **Cross Licenses:** | An arrangement between separate owners of patents to exchange patent rights. |
| **Declaration:** | An alternative method to a sworn statement in a patent application, which states that the signer is aware of the consequences of false statements. |

**Dependent Claim:**  A claim that refers to an earlier claim and adds detail to the earlier claim. This makes the claim narrower. *See* **Claim, Dependent.**

**Design Patent:**  A type of patent that protects an ornamental character of an object.

**Diligence:**  The activity of the inventor demonstrating steady progress toward reducing an invention to practice after conception. Diligence may be needed to be shown for the earliest inventor to prevail in an interference proceeding.

**Disclaimer:**  The renouncement of a patent claim.

**Doctrine of Equivalence:**  A technique used by patent holders to expand the literal language of their claims so that another who has made changes in the process or product to avoid infringement will still be an infringer.

**Double Patenting:**  Two or more patents or patent applications by the same inventor, claiming essentially the same invention.

**Duty of Disclosure:**  A requirement imposed by the Patent Office on all persons involved with a patent application, to disclose information to the Patent Examiner that may be material to the processing of the application.

**Effective Date:**  The date as of which a reference could be applied against an application. Also a date as of which an application or patent operates as such. A U.S. patent is effective as of its filing date as a reference against another U.S. application, and as a prior patent on its issue date.

**EPO:**  European Patent Organization. The organization of European countries that have organized to accept patent applications and issue patents valid in all member states.

**Examiner:**  An employee of the Patent and Trademark Office charged with determining the patentability of an invention.

**Example:**  A description, in detail, of one embodiment of an invention. The detail must be sufficient so that someone skilled in the art can understand and reproduce it.

**Ex Parte:**  Type of proceedings wherein there is no opposing party, as in normal patent application proceedings.

**Extension of Time:**  Permission granted by the Patent Office to extend the time period for the filing of a response or other action, with the filing of a petition and payment of the appropriate fee.

**Fee, Filing:**  A fee required by the Patent Office that must accompany the application for a patent.

**Fee, Issue:**  A fee required by the Patent Office for the conversion of an allowed application into a U.S. patent.

**File History:** Also known as Filewrapper. The complete file of a patent application and associated documents prepared by the applicant and the Patent Office personnel during the prosecution of the application.

**Filewrapper Estoppel:** A doctrine that prevents the patent owner from evaluating the claims of a patent more broadly than was understood when the patent was granted.

**Filing Date:** The date when the application reaches the Patent Office in complete form.

**First to File:** A system of patent laws, wherein the first application to file will be awarded the patent over all other inventors of the same invention.

**First to Invent:** A system of patent laws, followed by the U.S., wherein the first applicant to invent will be awarded the patent over all other inventors of the same invention.

**Harmonization:** The proposed alignment of the patent laws of all countries of the world to produce a system of patents that will be uniform in rights in all countries.

**Inducement to Infringe:** An act that causes or induces another to infringe a patent.

**Inequitable Conduct:** An act of improper conduct by an applicant, patentee, or a party of interest before the U.S. Patent and Trademark Office.

**Interference:** A proceeding of the Patent Office to determine who is the inventor when two or more applications or patents claim the same invention.

**Infringement:** The unauthorized making, using, or selling of a product or process that uses an invention protected by a patent.

**Interview:** A conference between the attorney of the applicant and the patent examiner in charge of the application. The conference can include the inventor and can be held in person or by phone.

**Inventor, First:** The person who files an application for a patent and was first to conceive and has reduced the invention to practice.

**Inventor, Joint:** One who in connection with one or more other persons conceives an invention and who plays a part in causing the same to be reduced to practice.

**Issue Fee:** Fee due when patent has been allowed, for its issuance.

**Know How:** A set of knowledge or technique not disclosed in a patent that will assist the operator of the patent in being able to practice the patent more efficiently.

**License:** The transfer of rights under a patent that does not amount to the transfer of all rights.

**Licensee:** The party obtaining rights under the terms of the license agreement.

**Licensor:**                    The party granted the rights under the terms of the license agreement.

**Machine:**                    An apparatus. One of the classes of subject matter that is patentable. They usually have two or more elements that cooperate with each other to achieve the purpose of the machine.

**Maintenance Fee:**                    The fees that must be paid to the Patent Office to maintain the patent grant. Payments are due at the 3½-, 7½-, and 11½-year periods from the date of the grant. *See also* **Annuity.**

*Manual of Classification:*                    The set of documents that lists all of the classes and subclasses that are contained in the U.S. Patent Classification System.

*Manual of Patent Examining Procedure:*                    Better known as the *MPEP*, the published set of rules and regulations for the examiners for dealing with the processing of patent applications. The manual can be purchased by the public.

**Marking, Patent:**                    The placing of one or more U.S. patent numbers on the surface of a patented article or on the surrounding packaging.

**New Matter:**                    Technical information added to a patent application beyond that which was originally present. It will be refused to be added because it was not present in the original application.

**Notice of Allowability:**                    A notice from the patent examiner that the patent application has been placed in a condition for allowance, and specific claims are allowable.

**Notice of Allowance:**                    A notice from the patent examiner that a patent will issue upon payment of the issue fee.

**Novelty:**                    A requirement for patentability. The invention claimed in the application must not have been known or used by others before invention by the applicant for the patent.

**Office Action:**                    A communication from the patent examiner to the applicant or his attorney, in which the examiner presents his opinions of the patent application.

*Official Gazette:*                    Also known as the *OG*. The weekly publication of the U.S. Patent and Trademark Office that lists the patents that were issued that week. Also lists official notices of the Office. A separate volume lists the trademarks issued that week.

**Paris Convention:** The common name given to the International Convention for the Protection of Intellectual Property, which was originally held in Paris. Countries that join this convention will give a patent application, submitted in one of their patent offices, the filing date of the earliest application in a member country, as long as it is filed within a 1-year period of the date of earliest filing.

**Patent Agent:** A technically trained person who is permitted to practice before the Patent Office in the interest of inventors.

**Patent Applied For:** Words used on an item, or associated with an item, to indicate that the item contains an invention, the patent for which has been applied for with the U.S. Patent Office. **Patent Pending** is another way of stating this.

**Patent Attorney:** A legally trained person, with a technical background, who is permitted to practice before the Patent Office in the interest of inventors.

**Patent Cooperation Treaty (PCT):** This treaty, of which the U.S. and Canada are members, provides a method for the preservation of rights in designated foreign countries for later filing of the patent application in those countries.

**Patent Pending:** *See* **Patent Applied For.**

**Patent Search:** Usually a search of the prior art to determine whether issued patents have disclosed the subject of the search.

**PDL:** Patent Depository Libraries, established throughout the U.S., which contain a file of patents for searching as well as trained personnel to help the searcher.

**Person Skilled in the Art:** A fictitious person who is supposed to have ordinary information and skill in the particular field of the invention.

**Petitions:** There are many specific types of petitions, but basically all are requests to the Commissioner of Patents to eliminate a wrong action taken by an examiner.

**Plant Patent:** A specific type of U.S. patent issued for a new asexually produced plant variety or species.

**Power of Attorney:** Authority given by the inventor to a patent agent or attorney to conduct all business with the U.S. Patent Office with respect to a specific patent application.

**Printed Publication:** Any document that has been produced and was made available to the public.

**Prior Art:** All prior knowledge, patent, literature, etc., relating to the claimed invention.

**Prior Art Statement:** A required statement by the patent attorney, agent, or inventor to the examiner of the references, patents, etc., considered to be material to the determination of patentability. This statement is required to be submitted within a brief period after the filing of the application.

**Priority Document:** A certified copy of an application in another country, submitted to the U.S. Patent Office to claim the filing date under the Paris Convention. *See* **Paris Convention**.

**Prosecution:** The process conducted between the applicant and the Patent Office concerning a patent application that results in its issuance as a patent, or as a rejection.

**Publication:** *See* **Printed Publication**.

**Public Sale:** The sale or offering to sell an item containing the invention to a member of the public.

**Reads On:** A term used in determining interference situations, wherein a claim includes, in its scope, certain subject matter of a claim in a different application.

**Reduction to Practice:** A term used to indicate that an invention has reached the stage that it has been implemented. **Actual reduction to practice** means that the invention has been conducted and reached a point where it has been physically confirmed. **Constructive reduction to practice** is the filing of a patent application showing to a person of average skill in the art how to make and use the invention.

**Reference:** Prior art cited by the examiner to support the rejection of claims for lack of novelty.

**Reissue Patent:** A patent that replaces an earlier patent that contained inadvertent errors in the claims, or seeks to reconsider the claims in view of newly found prior art.

**Rejection:** The refusal of the examiner to accept a claim because it lacks novelty or is obvious, and not patentable.

**Restriction:** A request from the examiner to limit the application to a single invention. This results when the examiner feels that the application contains more than one invention.

**Serial Number:** The number assigned to the patent application at the time of filing the application.

**Specification:** The written part of the patent application or the patent that describes the invention and how to use and make it.

**Statutory Period:** The time period in which a response must be submitted to an Office action, to avoid a holding of abandonment.

**Testimony:** Evidence presented by a witness under oath. This is distinguished from evidence derived from writings.

**Trademark:** A grant made by the U.S. Patent and Trademark Office to applicants to distinguish a product or service from the products or services of another.

**Trade Secret:** A method or practice that is kept secret by a person or organization for the purpose of obtaining a competitive advantage in the marketplace.

**Unobvious:**

An invention that would not be obvious to a person having ordinary skill in the art of the invention, assuming knowledge of all prior art.

**USC:**
Abbreviation for *United States Code.*

**Useful:**
Serving a purpose.

**Utility:**
Usefulness, a requirement of patentability.

**WIPO:**

The World Intellectual Property Organization, a part of the United Nations. It is a worldwide coordinating body for the improvement and harmonization of the patent laws.

**Working:**

A term applied to a patent in foreign countries which states that the patent must be "used" or "manufactured" within the boundaries of that country to remain valid and enforceable. Regulations vary widely from country to country as to what fulfills that condition.

# Index

# H

Hand search, 76, 89
Hopkins, Samuel, 7

# I

*Index to Classification*, 67
Individual inventor, 49
Infringement, 97–98
   definition, 97
   search, 68
INID, 57–59
Intellectual Property. 2. 4
Internet, 109–112
Interference, 115, 116,
Invention report, 42

# J

Jefferson, Thomas, 2

# K

Keyword, 77

# L

Laid open, 32
Legal documents, 97
Literature search, 77

# M

Maintenance fees, 49, 75
   definition, 75
   search, 75
*Manual of Patent Classification*, 67
Monopoly characteristic, 9

# N

North American Free Trade Association
   (NAFTA), 123
Non-Obviousness, 39
Notebooks, 113–7
Novelty requirements, 38

# O

Office action, 47
*Official Gazette (OG)*, 77

# P

Patent, 4,
   application, 43
   as technical literature, 43, 51–65
   contents, 52
   definition, 7,
   fees. 49
   first U. S. Patent, 7, 8,
   evolution, 41–50
   pending, 44, 47
   purpose, 5
   searching, 67–95
   specification, 43, 52–56
   term, 4
Patentability search, 68
Patent and Trademark Office (PTO), 13,
   26
   location, 25
Patent Cooperation Treaty (PCT), 120
Patent Depository Library (PDL), 63–65
Plant Patents, 14
Prior Art, 39
Process, definition of, 37
Property Characteristics, 9
Prosecution, 44
Provisional application, 48

# R

Record-keeping 41, 114–117
Reduction to Practice,
   actual, 41, 42
   constructive, 42
Romanian Patent, 62

# S

Serial number, 44
Slovakian patent, 61
Small business, fees for, 49
Specification, 43–44
Spelling differences, 63